MANAGEMENT OF DEPLETED URANIUM USED AS SHIELDING IN DISUSED RADIATION DEVICES

The following States are Members of the International Atomic Energy Agency:

AFGHANISTAN
ALBANIA
ALGERIA
ANGOLA
ANTIGUA AND BARBUDA
ARGENTINA
ARMENIA
AUSTRALIA
AUSTRIA
AZERBAIJAN
BAHAMAS
BAHRAIN
BANGLADESH
BARBADOS
BELARUS
BELGIUM
BELIZE
BENIN
BOLIVIA, PLURINATIONAL STATE OF
BOSNIA AND HERZEGOVINA
BOTSWANA
BRAZIL
BRUNEI DARUSSALAM
BULGARIA
BURKINA FASO
BURUNDI
CAMBODIA
CAMEROON
CANADA
CENTRAL AFRICAN REPUBLIC
CHAD
CHILE
CHINA
COLOMBIA
COMOROS
CONGO
COSTA RICA
CÔTE D'IVOIRE
CROATIA
CUBA
CYPRUS
CZECH REPUBLIC
DEMOCRATIC REPUBLIC OF THE CONGO
DENMARK
DJIBOUTI
DOMINICA
DOMINICAN REPUBLIC
ECUADOR
EGYPT
EL SALVADOR
ERITREA
ESTONIA
ESWATINI
ETHIOPIA
FIJI
FINLAND
FRANCE
GABON
GEORGIA
GERMANY
GHANA
GREECE
GRENADA
GUATEMALA
GUYANA
HAITI
HOLY SEE
HONDURAS
HUNGARY
ICELAND
INDIA
INDONESIA
IRAN, ISLAMIC REPUBLIC OF
IRAQ
IRELAND
ISRAEL
ITALY
JAMAICA
JAPAN
JORDAN
KAZAKHSTAN
KENYA
KOREA, REPUBLIC OF
KUWAIT
KYRGYZSTAN
LAO PEOPLE'S DEMOCRATIC REPUBLIC
LATVIA
LEBANON
LESOTHO
LIBERIA
LIBYA
LIECHTENSTEIN
LITHUANIA
LUXEMBOURG
MADAGASCAR
MALAWI
MALAYSIA
MALI
MALTA
MARSHALL ISLANDS
MAURITANIA
MAURITIUS
MEXICO
MONACO
MONGOLIA
MONTENEGRO
MOROCCO
MOZAMBIQUE
MYANMAR
NAMIBIA
NEPAL
NETHERLANDS
NEW ZEALAND
NICARAGUA
NIGER
NIGERIA
NORTH MACEDONIA
NORWAY
OMAN
PAKISTAN
PALAU
PANAMA
PAPUA NEW GUINEA
PARAGUAY
PERU
PHILIPPINES
POLAND
PORTUGAL
QATAR
REPUBLIC OF MOLDOVA
ROMANIA
RUSSIAN FEDERATION
RWANDA
SAINT KITTS AND NEVIS
SAINT LUCIA
SAINT VINCENT AND THE GRENADINES
SAMOA
SAN MARINO
SAUDI ARABIA
SENEGAL
SERBIA
SEYCHELLES
SIERRA LEONE
SINGAPORE
SLOVAKIA
SLOVENIA
SOUTH AFRICA
SPAIN
SRI LANKA
SUDAN
SWEDEN
SWITZERLAND
SYRIAN ARAB REPUBLIC
TAJIKISTAN
THAILAND
TOGO
TONGA
TRINIDAD AND TOBAGO
TUNISIA
TÜRKİYE
TURKMENISTAN
UGANDA
UKRAINE
UNITED ARAB EMIRATES
UNITED KINGDOM OF GREAT BRITAIN AND NORTHERN IRELAND
UNITED REPUBLIC OF TANZANIA
UNITED STATES OF AMERICA
URUGUAY
UZBEKISTAN
VANUATU
VENEZUELA, BOLIVARIAN REPUBLIC OF
VIET NAM
YEMEN
ZAMBIA
ZIMBABWE

The Agency's Statute was approved on 23 October 1956 by the Conference on the Statute of the IAEA held at United Nations Headquarters, New York; it entered into force on 29 July 1957. The Headquarters of the Agency are situated in Vienna. Its principal objective is "to accelerate and enlarge the contribution of atomic energy to peace, health and prosperity throughout the world".

IAEA NUCLEAR ENERGY SERIES No. NW-T-1.30

MANAGEMENT OF DEPLETED URANIUM USED AS SHIELDING IN DISUSED RADIATION DEVICES

INTERNATIONAL ATOMIC ENERGY AGENCY
VIENNA, 2023

Marketing and Sales Unit, Publishing Section
International Atomic Energy Agency
Vienna International Centre
PO Box 100
1400 Vienna, Austria
fax: +43 1 26007 22529
tel.: +43 1 2600 22417
email: sales.publications@iaea.org
www.iaea.org/publications

Printed by the IAEA in Austria
January 2023
STI/PUB/2020

IAEA Library Cataloguing in Publication Data

Names: International Atomic Energy Agency.
Title: Management of depleted uranium used as shielding in disused radiation devices / International Atomic Energy Agency.
Description: Vienna : International Atomic Energy Agency, 2023. | Series: IAEA nuclear energy series, ISSN 1995–7807 ; no. NR-T-1.30 | Includes bibliographical references.
Identifiers: IAEAL 22-01541 | ISBN 978–92–0–129122–6 (paperback : alk. paper) | ISBN 978–92–0–129222–3 (pdf) | ISBN 978–92–0–129322–0 (epub)
Subjects: LCSH: Depleted uranium. | Depleted uranium — Safety measures. | Depleted uranium — Management. | Radioactive waste disposal.
Classification: UDC 621.039.543.4 | STI/PUB/2020

FOREWORD

The IAEA's statutory role is to "seek to accelerate and enlarge the contribution of atomic energy to peace, health and prosperity throughout the world". Among other functions, the IAEA is authorized to "foster the exchange of scientific and technical information on peaceful uses of atomic energy". One way this is achieved is through a range of technical publications including the IAEA Nuclear Energy Series.

The IAEA Nuclear Energy Series comprises publications designed to further the use of nuclear technologies in support of sustainable development, to advance nuclear science and technology, catalyse innovation and build capacity to support the existing and expanded use of nuclear power and nuclear science applications. The publications include information covering all policy, technological and management aspects of the definition and implementation of activities involving the peaceful use of nuclear technology. While the guidance provided in IAEA Nuclear Energy Series publications does not constitute Member States' consensus, it has undergone internal peer review and been made available to Member States for comment prior to publication.

The IAEA safety standards establish fundamental principles, requirements and recommendations to ensure nuclear safety and serve as a global reference for protecting people and the environment from harmful effects of ionizing radiation.

When IAEA Nuclear Energy Series publications address safety, it is ensured that the IAEA safety standards are referred to as the current boundary conditions for the application of nuclear technology.

Sealed radioactive sources are used widely in industry, medicine, agriculture, mining, oil exploration and research. Most Member States have been using sealed sources for many decades. Many Member States do not have a sufficient or consistent framework for the management of the depleted uranium that is used for radiation shielding in sealed source equipment, containers and packages (in this publication, referred to collectively as devices).

This publication provides an overview of the devices that use depleted uranium for radiation shielding. It discusses methods for identifying those devices and containers, safeguards considerations and various options for the safe management of depleted uranium. Member States' experiences with depleted uranium management are provided in the annexes.

The IAEA is grateful to the five teams of international experts who contributed their experiences from in-field operations and helped draft this publication. The IAEA officer responsible for this publication was J.C. Benitez-Navarro of the Division of Nuclear Fuel Cycle and Waste Technology.

CONTENTS

1. INTRODUCTION

1.1. BACKGROUND

Depleted uranium (DU) arises as a by-product of the production of enriched uranium. It is defined as uranium containing a lower mass percentage of ^{235}U than is found in natural uranium [1, 2].

The widespread use of DU as a shielding material in both radiation devices and radioactive devices is a relevant issue for radioactive waste management because it is likely that in many cases the DU used in shielding sealed radioactive sources (SRSs) will have to be declared as radioactive waste. In particular, DU shielding is often used for [3–8]:

— High activity Category 1 and 2 gamma emitting sources, which are widely used in applications such as teletherapy devices, commercial irradiators and industrial radiography;
— Linear accelerators;
— Containers and packages designed for the storage, transport and disposal of high level waste (HLW) or spent nuclear fuel (SNF).

At the end of their operational life, both radiation and radioactive devices, especially those containing higher activity sources, result in residual radiation shielding materials such as DU, together with lead or tungsten. As a result, DU waste may be generated. Each Member State should establish a decision making process for designating a disused source as radioactive waste, taking into account the potential effects of such a designation on subsequent management options [9]. In such circumstances, when devices containing DU are declared as radioactive waste, the safety requirements in the publications Predisposal Management of Radioactive Waste, IAEA Safety Standards Series No. GSR Part 5 [10], and Disposal of Radioactive Waste, IAEA Specific Safety Requirements No. SSR-5 [11], will apply.

DU shields can be safely and securely managed alongside disused sealed radioactive sources (DSRSs) and radioactive waste in the same facilities and by the same staff. Some specific considerations are set out in this publication and can readily be accomplished.

Radioactive devices removed from operation may contain their radioactive sources, or these sources may be removed. If these SRSs are at the end of their useful life, they are defined as 'spent' or 'disused' (i.e. DSRSs). For sake of brevity, only the term 'disused' is employed in the rest of this publication.

In this publication, a 'radioactive device' is a device that holds and shields an SRS for use in its given application, such as teletherapy. Storage and transport containers, which may use DU to shield SRS and other radioactive items, are also included within this definition.

Currently, a large inventory of DSRSs exists that has accumulated in various Member States, and it is likely to continue increasing in the near term, given the current and future potential use of SRSs worldwide [3–8]. As a result of current issues associated with the safe management of DSRSs and the control of nuclear material, DU management is a topic that needs proper attention.

The IAEA Secretariat and its Member States have taken steps to lower the risks associated with DSRSs, including the establishment of the Code of Conduct on the Safety and Security of Radioactive Sources [12] and its supplementary guidance on the management of DSRSs [9]. Simultaneously, a binding international regime for the safety of radioactive waste management and spent fuel (the Joint Convention) has been adopted [13].

This publication supplements existing IAEA reports on the safe management of DSRSs and nuclear material, including their disposal [14, 15].

The management of DU contained in radiation devices and radioactive devices once their DSRSs have been removed has not been addressed to date in a comprehensive and systematic manner. The need for such a publication has been discussed and highlighted at various IAEA Consultants' Meetings and Technical Meetings, as well as at regional events, such as Regional Coordination Meetings, Workshops and Training Courses of the IAEA's Technical Cooperation Programme.

Given this background, it is timely and important from the IAEA perspective that a publication be prepared on this issue, focusing on the various aspects related to the management of DU in disused devices within Member States. Considering that this topic has not been addressed in the IAEA SRS programme to date, it is anticipated that the publication will provide the much needed information required by Member States for the management of DU shields associated with disused radioactive devices, as well as radiation devices.

The publication will be of direct relevance to policy makers, operators and regulators in Member States that are exploring options, or developing and implementing strategies, for the safe management of disused devices.

1.2. OBJECTIVE

The objective of this publication is to provide information on:

— Methods to identify devices containing DU used as shielding;
— The hazards of DU found in such devices;
— The safe handling of DU in such devices;
— Various options for the management of DU shields;
— Safety, security and safeguards considerations for the control and traceability of DU arising from disused devices, based on international experience.

Another key objective of this publication is to raise international awareness of this emerging field of interest. Guidance and recommendations provided here in relation to identified good practices, represent experts' opinions but are not made on the basis of a consensus of all Member States.

1.3. SCOPE

Since this is an emerging field of interest, this publication only provides an overview of the field and experiences in some Member States; it does not provide guidance and recommendations on specific aspects of DU management. However, some general recommendations are provided to address short term issues.

This publication focuses only on DU from disused devices at the end of their operational life. The scope of the report primarily covers medical and industrial devices, in particular those containing higher activity, gamma emitting SRSs that require radiation shielding or collimation. Containers and packages that contain DU alloys and are designed for the storage and transport of gamma emitting SRSs are also discussed.

The publication presents relevant information on technical issues and factors and specific Member State experiences (see the annexes) leading to the identification of potential options for the management of DU shields. Various options for safe, secure and cost effective solutions have been explored, ranging from returning to manufacturer, reuse, recycling and storage to disposal in licensed facilities. The handling of DU as exempted or cleared material, which can be released without specific radiological controls (unrestricted release), is not an option considered in this publication.

The safeguards control of DU used as shielding material is addressed, referring to the IAEA Safeguards Implementation Guide for States with Small Quantities Protocols [16].

1.4. STRUCTURE

This report contains eight sections, three appendices and 24 annexes (with the addition of an introductory section). The 24 annexes summarize national experiences with radiation and radioactive devices that contain DU.

Section 2 presents an overview of the characteristics of uranium and DU. In addition, details such as radioactive half-lives and specific activities of uranium isotopes, the uranium decay series and additional details on the general, chemical and radiological characteristics of DU are provided. Notably, the lower radioactivity of DU compared with natural uranium, as well as its high density, which makes it an attractive option as a radiation shielding material, are described.

Section 3 describes the uses of DU in radiation and radioactive devices and provides an overview of these devices. To assist the reader, a variety of photographs and schematics are provided to illustrate how DU is incorporated into these devices.

Section 4 describes how DU-containing devices can be identified. Again, to assist the reader, a variety of photographs are provided that illustrate device and component labelling, as well as example methods for identification of DU devices if labelling and documentation are inadequate or missing.

Section 5 presents safeguards considerations for DU shielded devices, namely obligations, responsibilities and steps to take when dealing with DU shields.

Section 6 details the safety and security factors that need to be considered when handling, transporting and/or storing DU. These include nuclear, industrial and radiological safety and security considerations for transport and storage.

Section 7 describes the options for managing DU — that is, returning to manufacturer/supplier, reusing or recycling, or storage and disposal. Options are described in the context of both Member States that have large nuclear programmes and those that only have DU from shielded devices. General information and inventory record keeping are also described.

Section 8 summarizes this report and presents the steps forward.

Appendix I describes the general aspects of radioactive waste.

Appendix II provides a case study of DU shielding control and protection (France).

Appendix III presents specific regulatory requirements for the storage and transport of DU (Hungary).

The annexes comprise 24 national reports that were prepared by Member State representatives in an IAEA Technical Meeting (19–23 August 2019, Vienna, Austria) according to a brief questionnaire that was provided by one of the meeting's consultants. These national reports provide a snapshot of the status of DU management in the Member States.

2. CHARACTERISTICS OF URANIUM AND DU

2.1. CHARACTERISTICS OF URANIUM

Uranium (U) is a naturally occurring radioactive element. In its pure form, it is a silver-coloured heavy metal, similar to lead, cadmium and tungsten. Like tungsten it is very dense; at about 19 g per cubic centimetre, it is 70% denser than lead. It is so dense that a small 10 cm cube would weigh 20 kg.

Natural uranium contains three main isotopes: ^{238}U (99.27% by mass), ^{235}U (0.72% by mass) and ^{234}U (0.0054% by mass). All three of these isotopes emit alpha particles as their primary radiation. The following provides a summary of the radiological properties of uranium isotopes and their decay products [17].

When radionuclides in a decay series have long half-lives, such as ^{238}U, ^{234}U and ^{230}Th, the resulting in-growth of decay products further along the decay chain, for example, ^{226}Ra, occurs very slowly (see TABLE 1). Thus, the abundance of decay products like ^{226}Ra will be insufficient to produce a significant radiological hazard for tens of thousands of years. Therefore, the only radionuclides that occur in sufficient abundance to have an impact on radiological hazards are the shorter lived isotopes: ^{234}Th and ^{234m}Pa from ^{238}U, and ^{231}Th from ^{235}U [17]. Within a few months following production of DU, these isotopes will have built up to their maximum concentration, reaching secular equilibrium with their parent radionuclide ^{238}U.

Uranium enrichment facilities concentrate ^{235}U above the level of natural uranium. The degree of enrichment depends on the uranium's potential use: between 2% and 5% for low enriched uranium for nuclear power reactors, and up to 20% for research reactors. The by-product of enrichment is DU.

DU typically contains ^{238}U (99.75% by mass) and relatively smaller amounts of ^{235}U (0.25% by mass) and ^{234}U (0.002% by mass). Since ^{238}U has a much longer half-life relative to the lighter uranium isotopes ^{234}U and ^{235}U, DU emits about 60% less alpha radiation than natural uranium (on a mass basis) because the more radioactive, shorter lived uranium isotopes have been partially removed.

Because of its very long half-life, ^{238}U has little radioactivity per gram. In contrast, ^{226}Ra, with a half-life of 1602 y, has a much higher (by orders of magnitude) specific activity of 37 GBq/g. The specific activities of various mixtures of uranium isotopes are presented in Table 2 [17].

The calculated specific activity of natural uranium is approximately 2.59E+4 Bq/g (0.70 μCi/g), whereas the specific activity of DU is approximately 1.48E+4 Bq/g (0.4 μCi/g), primarily because of the removal of ^{234}U, which contributes to approximately 49% of the total activity of natural uranium [17–19].

Figure 1 shows the isotopic profile for ^{238}U decay chain in natural uranium ore prior to refining at a mine site. Figure 2 shows the isotopic profile for ^{238}U decay chain after the refining process. Refining removes all decay products of ^{238}U and ^{235}U which are present at the extraction time. Enrichment generates uranium with increased concentrations of ^{235}U and ^{234}U and DU remains with higher ^{238}U concentration, compared to those in natural

uranium.

TABLE 1. RADIOLOGICAL PROPERTIES OF URANIUM ISOTOPES AND DECAY PRODUCTS

Radionuclide	Half-life	Principal types of radiation
Uranium isotopes		
U-238	4.5E+9 a[a]	alpha
U-235	7.1E+8 a	alpha, gamma
U-234	2.5E+5 a	alpha
Decay products		
Th-234 (from U-238)	24.1 d	beta, alpha
Th-231 (from U-235)	25.5 h	beta, alpha
Th-230 (from U-234)	8.0E+4 a	alpha, gamma
mPa-234 (from U-238)	1.2 min	beta, gamma

[a] a: year

TABLE 2. SPECIFIC ACTIVITIES OF URANIUM

Uranium isotope mixture	% U-235	Specific activity (Bq/g)
Pure U-238	0	1.22E+4
Depleted	0.20	1.48E+4
Natural	0.72	2.59E+4
Enriched	2.0	3.70E+4
Enriched	20	3.33E+5

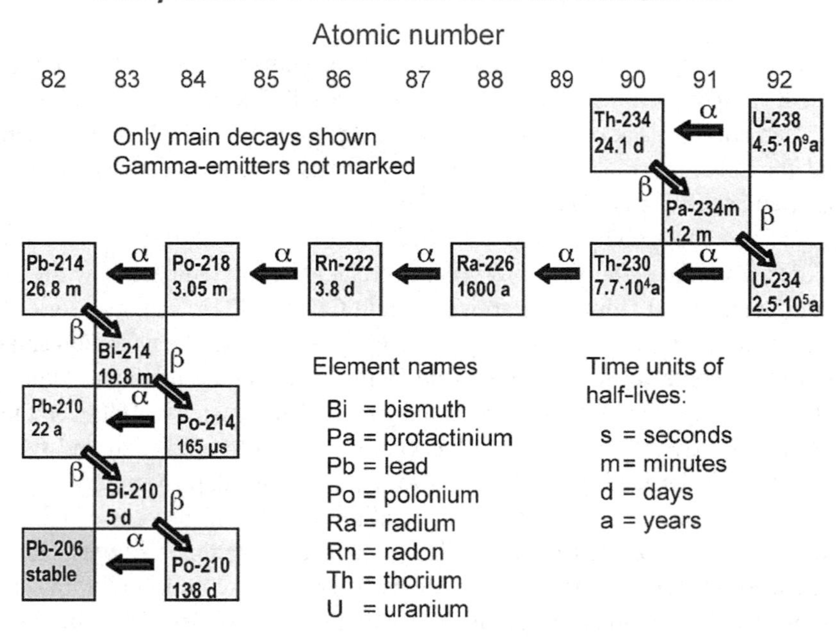

FIG. 1. Isotopic profile of ^{238}U in uranium ore prior to refining (courtesy of L. Pillette-Cousin, France).

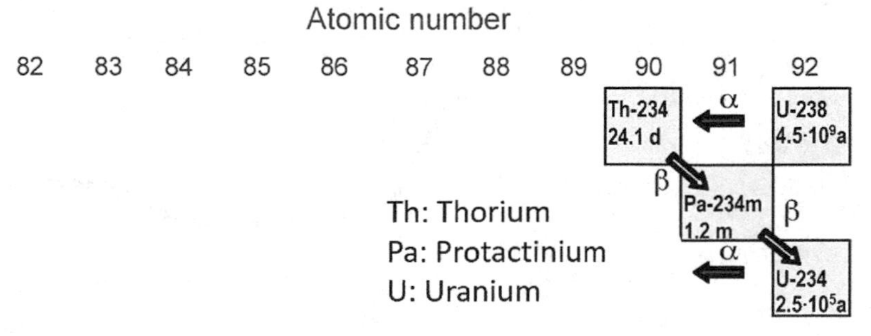

FIG. 2. Isotopic profile of ^{238}U in DU after enrichment of natural U (courtesy of L. Pillette-Cousin, France).

2.2. CHARACTERISTICS OF DU

As mentioned in the previous section, DU is a by-product of the process to enrich natural uranium, and contains approximately 99.75% ^{238}U, 0.25% ^{235}U and 0.002% ^{234}U. That is, DU has a much lower content of the high energy, fissile isotope ^{235}U relative to natural uranium.

DU is considerably less radioactive than natural uranium, not only because it has less ^{234}U and ^{235}U per unit mass than does natural uranium, but also because essentially, all traces of decay products beyond ^{234}U and ^{231}Th have been removed during the extraction and chemical processing of the uranium prior to enrichment. The specific activity of uranium alone in DU is 14.8 Bq/mg, compared with 25.4 Bq/mg for natural uranium.

2.2.1. General characteristics of DU

Metallic DU is among the densest materials known, with a density of 19.1 g/cm^3. It is a shiny silver-gold metal. It is pyrophoric as chips or powders. It readily oxidizes upon exposure to air to form a layer of yellow-green and/or black oxide on the surface. Because DU is 1.67 times as dense as lead, it is a valuable material for radiation shielding. Furthermore, DU has been used extensively in a wide variety of other applications (ballast, counterweights).

Figure 3 illustrates the activity of DU as a function of time relative to its initial activity [20]. For comparison purposes, a similar activity ratio for a commercial low level radioactive waste facility is provided, based on data from the disposal facility at Barnwell, South Carolina [21]. The inventory of such a low level waste (LLW), near surface disposal facility comprises mainly waste from the operation and outage of nuclear power plants, as well as some waste from fuel cycle facilities containing uranium. Although the activity in the commercial LLW disposal facility would decrease by more than a factor of 100 over a few hundred years, because of the decay of ^{60}Co, ^{137}Cs, ^{90}Sr and other major short lived activation and fission radionuclides, the activity of uranium in the facility would begin to increase after approximately 1000 years. This is to be considered when planning the disposal of large quantities of DU. In the case of DU from radiation and radioactive devices, the quantities of DU that could be managed as radioactive waste are much lower than the quantities generated by uranium enrichment facilities. Therefore, the impact of a low amount of DU from radiation and radioactive devices on the long term radioactivity in a LLW disposal facility would be negligible.

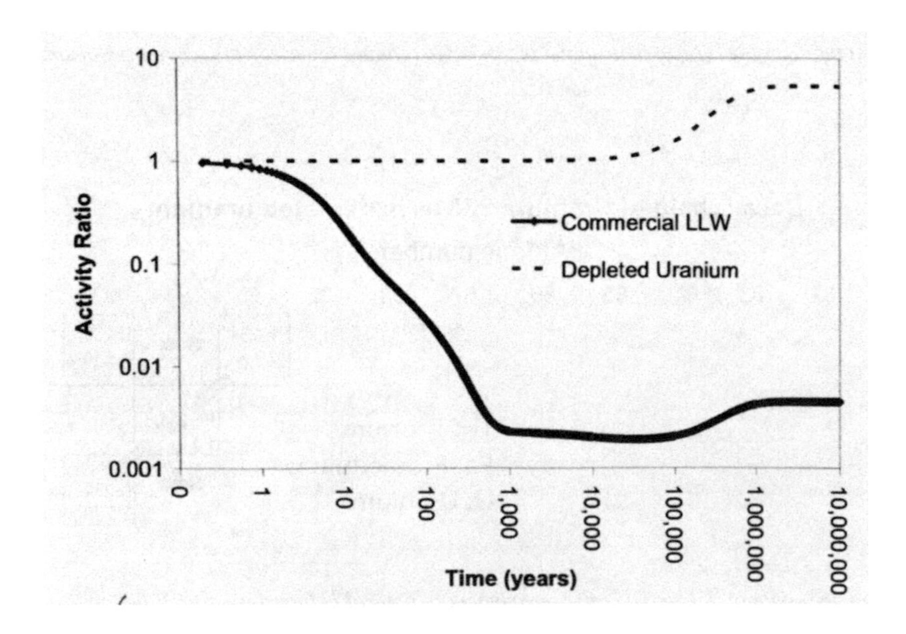

FIG. 3. Ratio of the activities of commercial LLW and DU as a function of time, relative to their initial activities (reproduced from Ref. [20] with permission).

2.2.2. Chemical characteristics of DU

The main biological risk associated with DU is not its radioactivity, but its chemical toxicity [17–22]. Finely divided DU, such as powder, can be harmful if ingested or inhaled because of its chemical toxicity. Like mercury, cadmium and other heavy metal ions, excess uranyl ions can affect the kidneys. High concentrations in the kidney can cause damage and, in extreme cases, kidney failure. The general medical and scientific consensus is that in cases of high intake, DU is likely to become a chemical toxicology problem before it is a radiological problem. Since DU is mildly radioactive, once inside the body it also irradiates organs, but the primary health effect is associated with its chemical action on body functions [19].

Solid metal DU components are stable and present little risk or toxic hazard. DU in the form of chips or powders can readily ignite (i.e. it is pyrophoric) and is difficult to extinguish. Thick layers of oxides may form during DU storage in humid conditions, which can be inhaled or ingested.

2.2.3. Radiological characteristics of DU

As mentioned regarding chemical hazards, DU presents a contamination hazard when it is in the form of fine powder or corrosion. Because it is weakly radioactive as a result of alpha particle emissions, DU is usually considered to be more of an internal radiological hazard than an external radiation hazard.

Radiation Protection and Safety of Radiation Sources: International Basic Safety Standards, IAEA Safety Standards Series No. GSR Part 3 [23], sets limits for occupational exposure and public exposure, which are based on the recommendations of the International Commission on Radiological Protection (ICRP). Requirement 12 of GSR Part 3 [23] states: "[t]he government or the regulatory body shall establish dose limits for occupational exposure and public exposure, and registrants and licensees shall apply these limits." The proposed values of the dose limits for public and occupational exposure are given in Schedule III of GSR Part 3 [23]. These dose limits apply to the sum of the doses from external exposure in the specified period and the relevant committed doses from intakes in the same period. Therefore, the relevant dose from exposure to any combination of uranium isotopes, including those in DU, is considered. Uranium-238 radiotoxicity is low but cannot be neglected, particularly in the case of inhalation: according to GSR Part 3, committed effective dose factor per becquerel via inhalation may be up to 7.3E–6 Sv/Bq.

Appendix I provides additional information on the general aspects of radioactive waste classification.

3. USES OF DU

3.1. DEVICES THAT CONTAIN DU

Radiation and radioactive devices generally incorporate sufficient shielding to absorb radiation to a level that is harmless to workers and the public. Commonly used shielding materials consist of one or more of the following materials: DU, tungsten and lead. Collimators are designed to direct and focus radiation beams in radiography or teletherapy, thereby reducing radiation levels and subsequent doses outside the collimated beam [24]. Some devices use DU for collimation. Relative to radioactive devices, only a small number of radiation devices, such as linear accelerators, have used DU for shielding and collimation [25].

A detailed list of radioactive devices and their models that contain DU is provided in the Safeguards Implementation Guide for States with Small Quantities Protocol, IAEA Services Series No. 22 [16].

References [5] and [8] provide a summary of devices that may have contained higher activity sealed sources used extensively worldwide.

TABLE 3 shows the functional applications and radionuclides of interest for some devices. Some descriptions of radioactive devices follow.

Medical teletherapy devices contain a high activity SRS, typically ^{60}Co, or, less commonly, ^{137}Cs, with typical activity levels in the range of 148 TBq to 370 TBq (4000 Ci to 10 000 Ci), or even higher in some cases [26]. The SRS is placed inside the teletherapy head, which usually contains large amounts of high density shielding material, such as DU, tungsten or lead. Teletherapy devices may contain smaller amounts of ancillary components, such as the drawer, wheel, source holder and trimmer bars (see Figs 4–9). Examples of brachytherapy devices are shown in Figs 10 and 11.

Most self-shielded irradiators, which may contain DU as a shielding material, have one or more sealed sources containing either ^{60}Co or ^{137}Cs (see Fig. 12).

In some self-shielded irradiators from J.L. Shepherd & Associates' Model 109 series, a design feature is the incorporation of plated DU shielding, fixed in the irradiator body (see Fig. 13).

Blood/tissue irradiators contain ^{60}Co or ^{137}Cs (see Fig. 14).

TABLE 3. FUNCTIONAL APPLICATIONS AND RADIONUCLIDES OF INTEREST FOR SOME RADIOACTIVE DEVICES

Devices/applications	Radionuclide(s)
Teletherapy devices	
Medical/cancer therapy	Co-60, Cs-137
Multibeam (gamma knife)	Co-60
Brachytherapy devices	Co-60, Cs-137, Ir-192
Self-shielded irradiators	Co-60, Cs-137
Industrial radiography	Co-60, Se-75, Cs-137, Yb-169, Tm-170, Ir-192
Measurement gauges	Co-60, Cs-137, Ir-192, Am-241
Well logging devices	Co-60, Cs-137

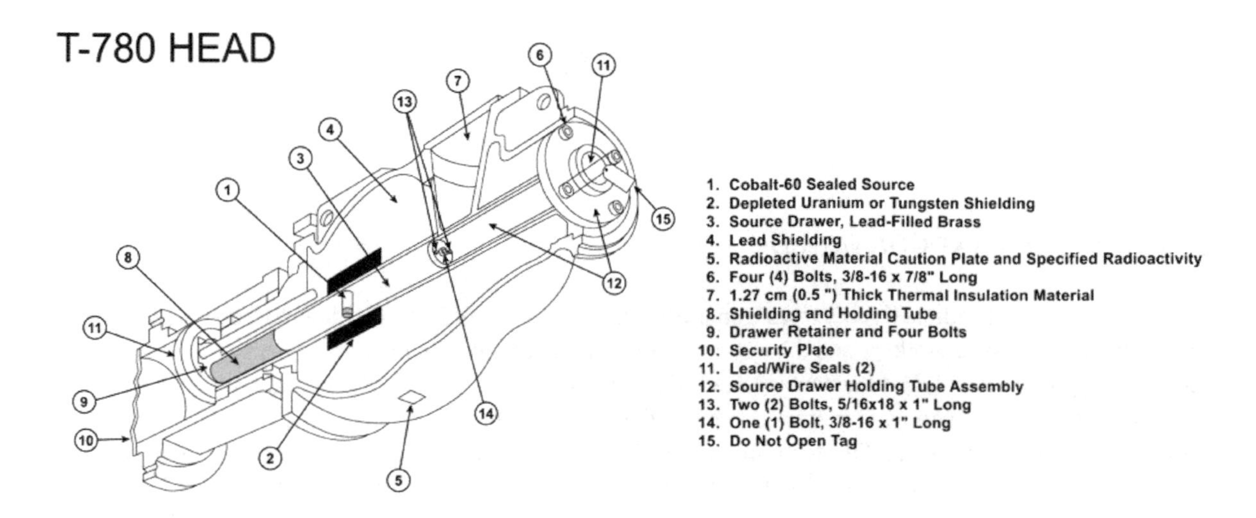

FIG. 4. Example of a teletherapy device head. The DU shielding (labelled as no. 2 in the figure) is surrounding the radioactive source (courtesy of Best Theratronics).

(a)

(b)

FIG. 5. An empty teletherapy unit, model PICKER C8M80, with (a) DU in the head and (b) the shutter mechanism.

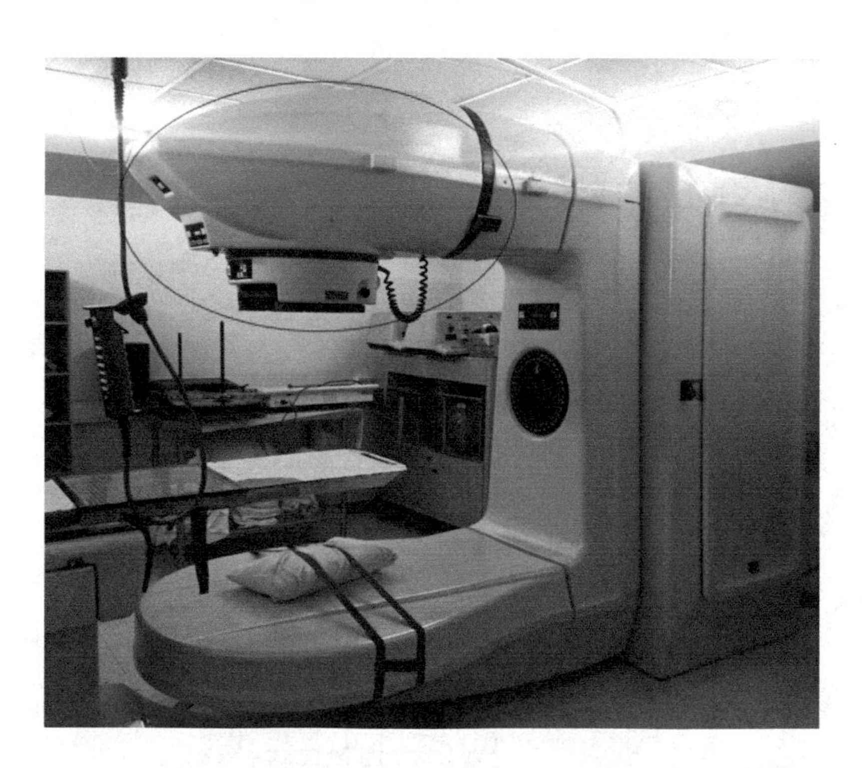

FIG. 6. Photograph of a disused medical teletherapy device, showing where the head and collimators are located (circled in red).

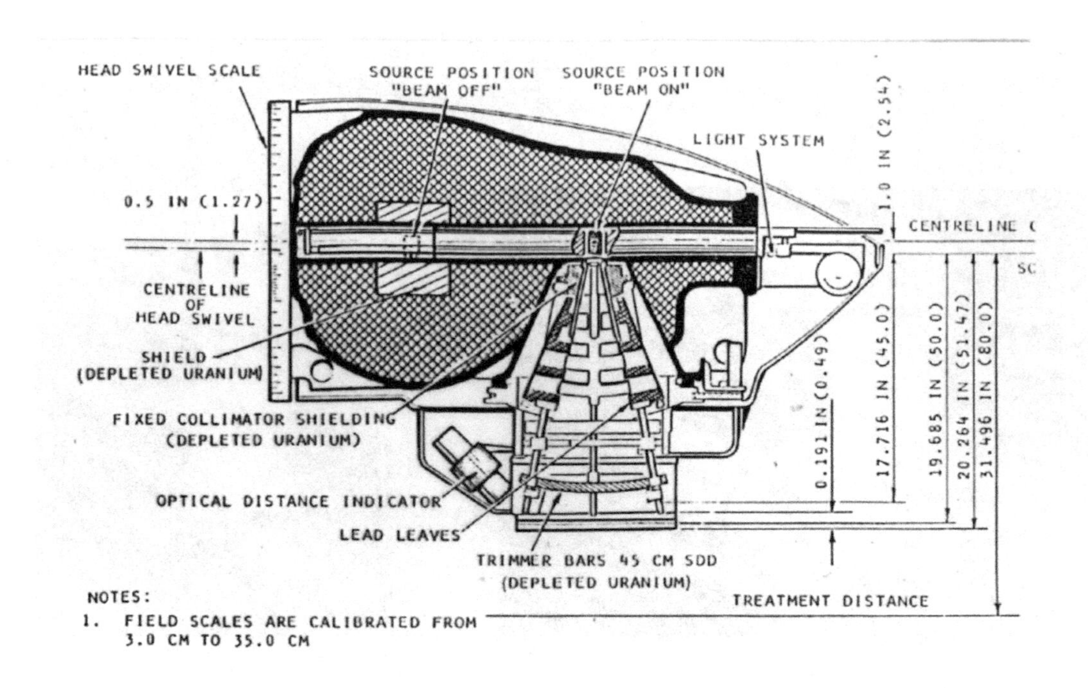

FIG. 7. Example of a teletherapy head showing DU components for shielding and collimation (courtesy of Best Theratronics).

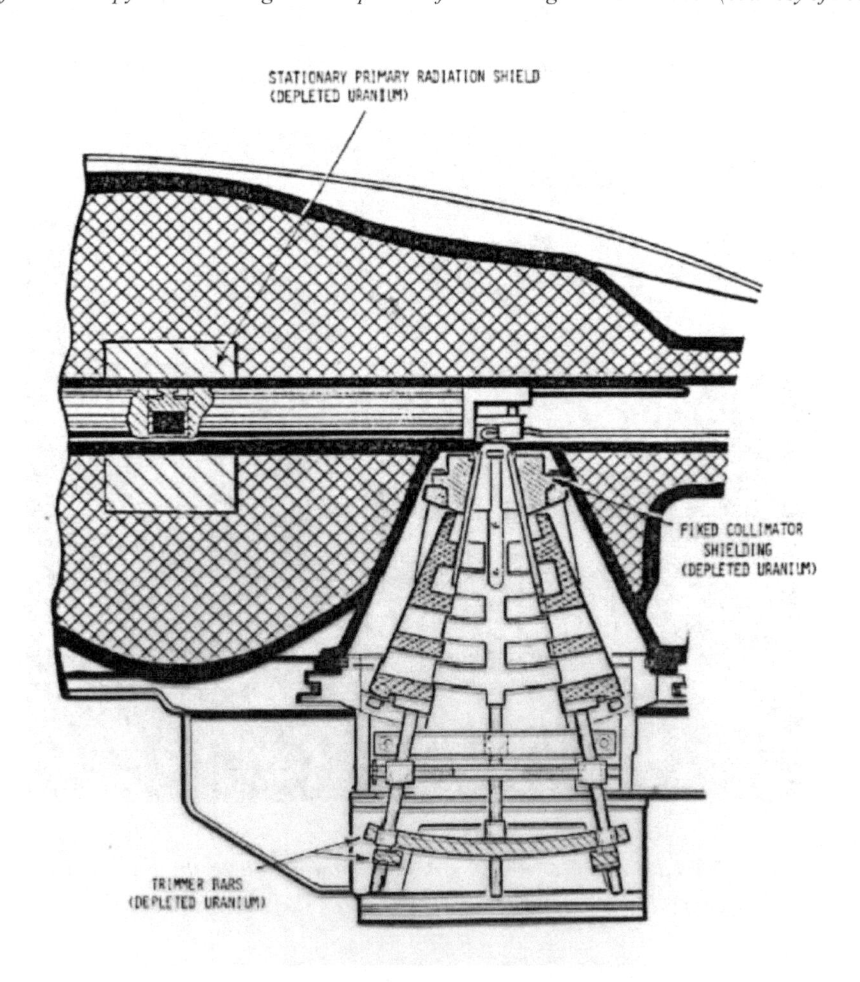

FIG. 8. Detailed view of DU collimators on a teletherapy device (courtesy of Best Theratronics).

3.1.1. Teletherapy devices

Figure 9 shows examples of a DU collimator (trimmer bars) used in a teletherapy device.

3.1.2. Brachytherapy devices

Brachytherapy is a form of radiotherapy where one or several SRSs are placed inside or next to a tumour requiring treatment. From the medical standpoint, there are three levels of brachytherapy dose, which determine the duration of treatment: low dose rate (LDR), medium dose rate (MDR) and high dose rate (HDR).

LDR brachytherapy needs several low activity sources, and equipment is composed of an afterloader from the treatment itself and a source changer and storage unit. Both MDR and HDR brachytherapy use only one source, and the same device (the afterloader) plays the role of storage and treatment unit. Many devices for LDR and MDR/HDR brachytherapy have used DU as primary shielding (Figs 10 and 11).

3.1.3. Self-shielded irradiators

A self-shielded irradiator is defined by the American National Standards Institute as

> "An irradiator in which the sealed source(s) is completely contained in a dry container constructed of solid materials, the sealed source(s) is shielded at all times, and human access to the sealed source(s) and the volume(s) undergoing irradiation is not physically possible in its designed configuration" [27].

Self-shielded irradiators are used mostly for blood irradiation, biomedical and radiation research and calibration of other devices. Blood banks irradiate blood and cellular blood components with 25 Gy to 35 Gy to prevent at-risk patients from developing graft versus host disease. See Figs 12–14.

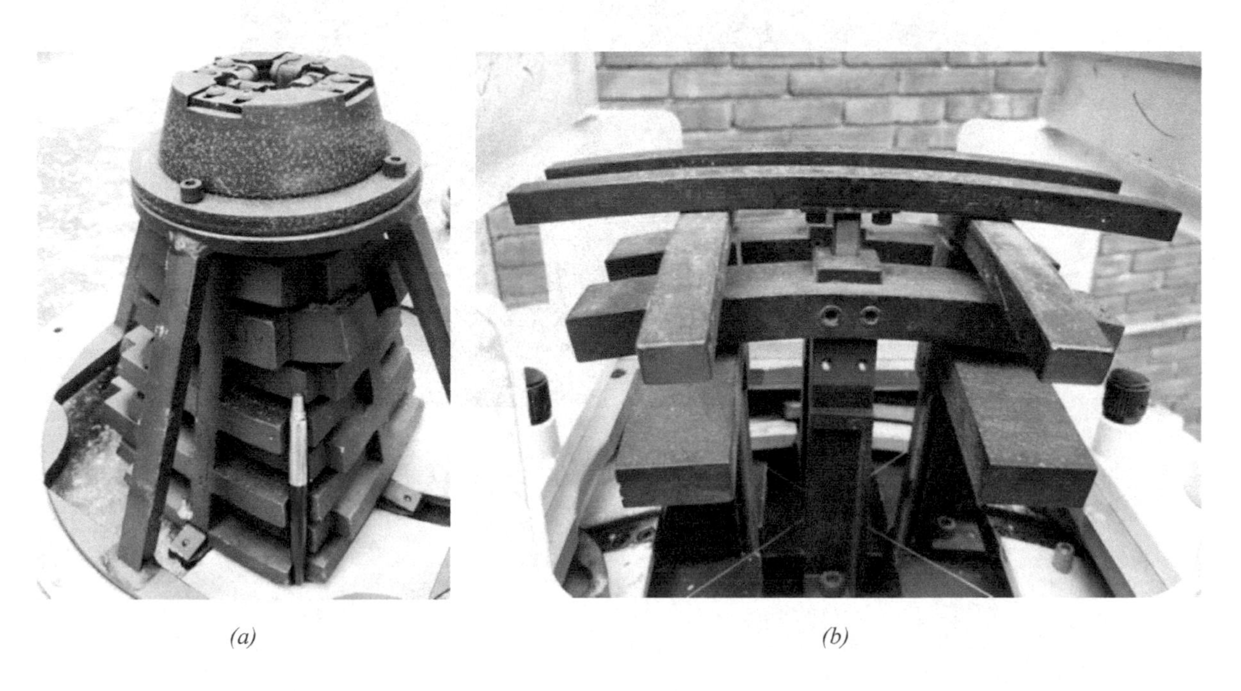

(a) *(b)*

FIG. 9. Examples of assembled collimator (a) and trimmer bars (b) with DU.

FIG. 10. LDR brachytherapy source changer and storage unit (left of image) and afterloader (centre of image) with DU shielding (courtesy of CIS Bio International, France).

FIG. 11. HDR brachytherapy unit with DU shielding (courtesy of L. Pillette-Cousin, France).

(a) *(b)* *(c)*

FIG. 12. Self-shielded irradiator model AGAT-VUI with DU from the former USSR, with (a) front view, (b) back view and (c) identification of DU shielding.

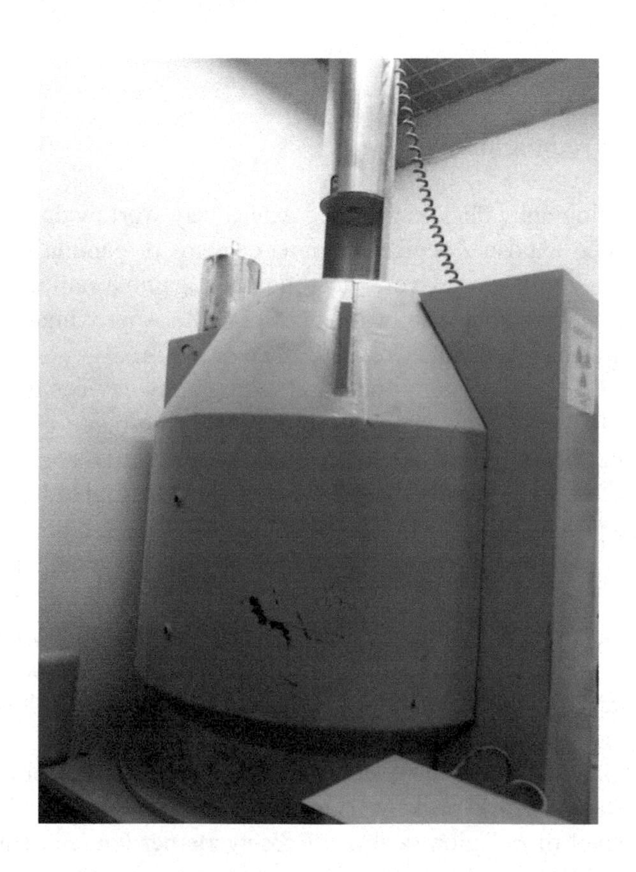

FIG. 13. Model 109 self-shielded irradiator from J.L. Shepherd & Associates.

FIG. 14. Blood/tissue irradiator (courtesy of CIS Bio International, France).

3.1.4. Industrial gamma radiography

Industrial gamma radiography devices are used extensively worldwide for inspection of pipelines, welds and so on. Also called exposure devices, gamma cameras or gamma projectors, they commonly take the form of portable radiography devices. Most industrial gamma radiography devices consist of a shield, which may contain DU, and one or more sources [24, 28]. Approximately 80% of those currently in use contain ^{192}Ir; they may also contain ^{60}Co, ^{75}Se, ^{169}Yb or ^{170}Tm.

The reason why the shielding of many industrial radiography devices incorporates DU is that this material is much denser than lead. This enables the devices to be made physically smaller and more portable than would be possible with lead shielding alone. The use of DU as shielding results in devices that are relatively heavy, often weighing 15 kg or more. Typical industrial radiography sources contain up to approximately 3.7 TBq (100 Ci) of ^{192}Ir [3]. Because these sources are housed in portable equipment, they can pose a potential safety and security risk. A schematic drawing of the SPEC 150 portable radiography device is presented in Fig. 15, with an S-channel showing the various components of the device, including the DU shield surrounding the S-tube through which the source assembly travels.

An image of the front and back of the SPEC 150 radiography device is shown in Fig. 16.

A photo of a GammaMat radiography device with a straight channel is shown in Fig. 17.

Another old and disused radiography device is shown in Fig. 18.

FIG. 20 show containers used for the exchange of industrial radiography sources. The containers typically use lead or DU shielding. DU packages have to be shipped as radioactive, even when they carry no sources, since a low level of radiation (a few microsieverts per hour) is emitted from the DU even when no source is present [29].

In industrial gamma radiography, shielding reduces both the size of the controlled area and the radiation dose received by radiographers. Shielding in the form of collimators is designed so that the radiation beam is primarily in the direction necessary for radiography. Collimators are made from radiation absorbent material such as lead, DU or tungsten, and are available in panoramic and directional formats (see Figs 21 and 22).

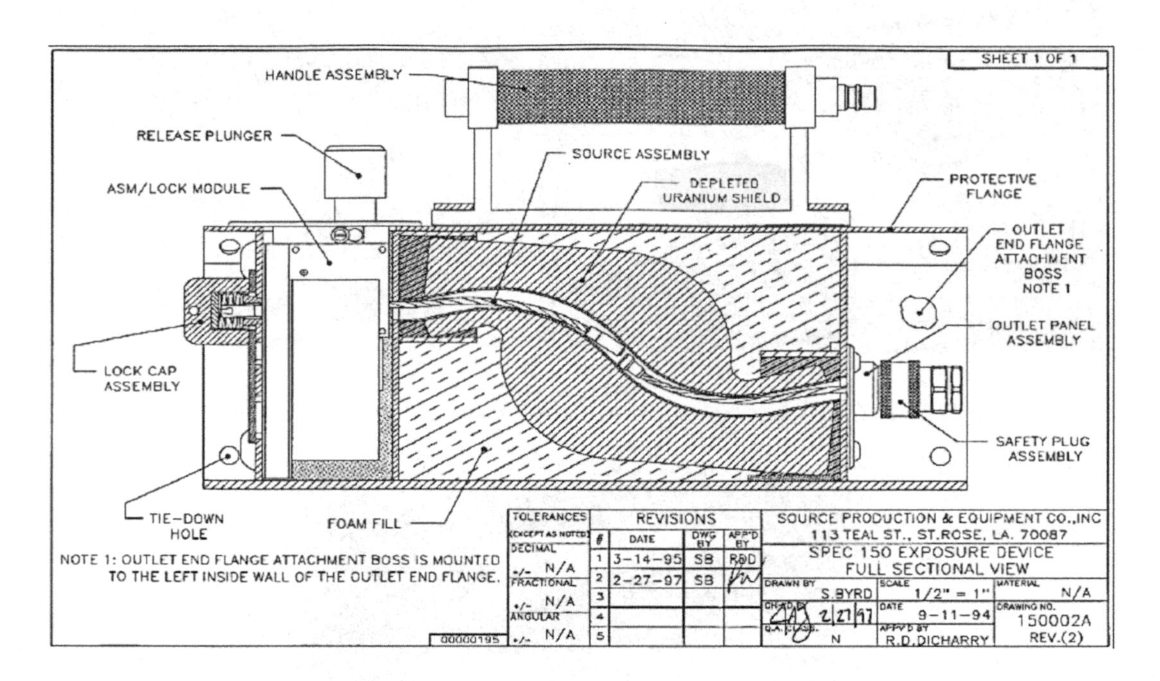

FIG. 15. Schematic of a SPEC 150 model portable radiography device, showing the various components of the device, including the DU shield (courtesy of Source Production & Equipment Co.).

FIG. 16. Photograph of the front and back of a SPEC 150 device (courtesy of Source Production & Equipment Co.).

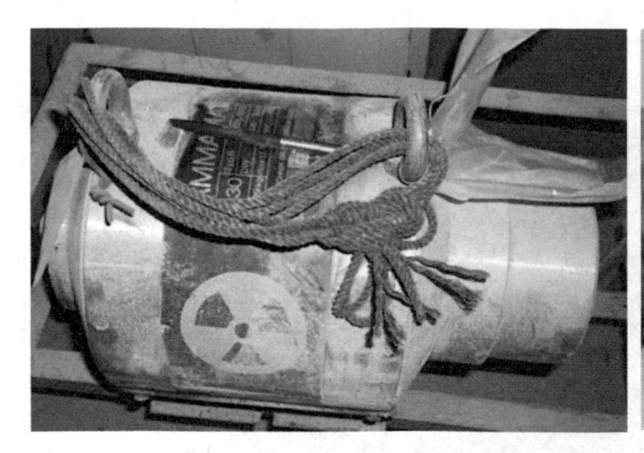

FIG. 17. GammaMat radiography projectors, Model TK30, with DU shielding.

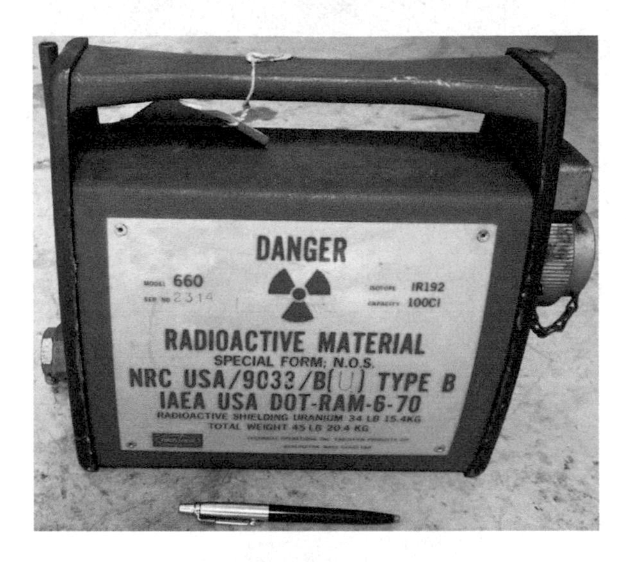

FIG. 18. TechOps model 660 radiography projector containing DU shielding.

FIG. 19. Sentinel 650L radiography device source changer (courtesy of QSA Global, Inc.).

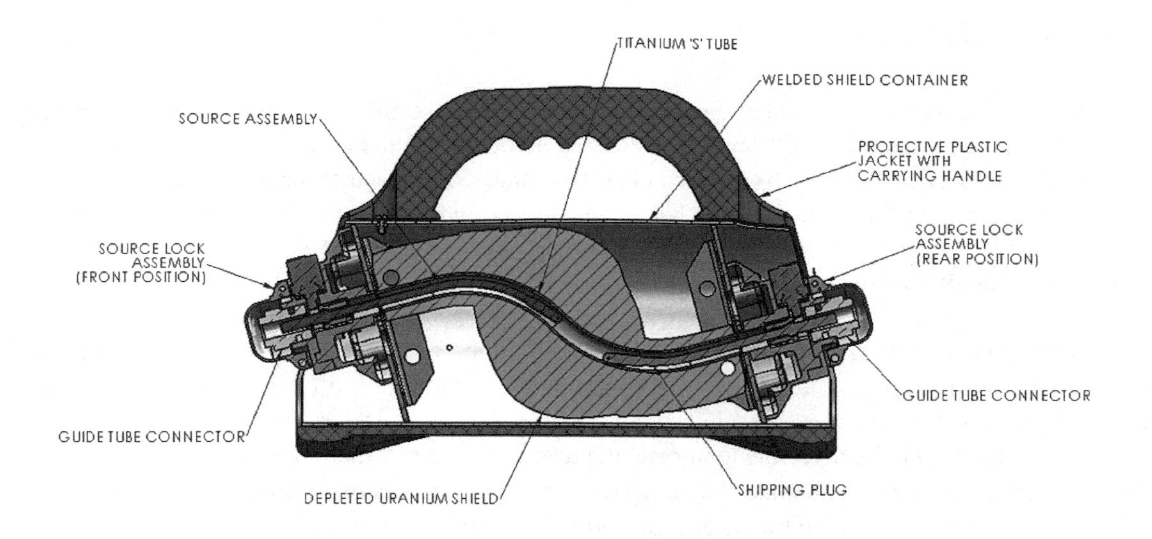

FIG. 20. Sentinel 880SC radiography device source changer (courtesy of QSA Global, Inc.).

FIG. 21. Directional collimator for ^{192}Ir industrial gamma radiography, containing DU (courtesy of Actemium NDT-P&S — VINCI Energies).

FIG. 22. Panoramic collimator for ^{192}Ir industrial gamma radiography, containing DU (courtesy of Actemium NDT-P&S — VINCI Energies).

3.1.5. Measurement gauges

Radioactive gauges (or nuclear gauges) are devices that use SRSs to measure parameters such as thickness, density, moisture or fill level in processes involving (a) materials with extreme temperatures or harmful chemical properties; (b) materials that are damaged by contact; or (c) packaged products. As radiation is emitted all the time, shielding has to be ensured, and sometimes DU is used (see Figs 23–25).

3.1.6. Well logging devices

In the 1980s, well logging tools began using radiation shielding DU cylinders when performing gamma density diagraphy. The tool's ^{137}Cs source emits gamma rays in a geological formation, and scattered gamma rays are measured by sodium iodide detectors. DU shielding cylinders are placed between the source and the detectors to impede the detection of direct gamma rays emitted by the source. Since the 1990s, DU has been replaced by tungsten for this use, but many disused DU shields of this kind may still be found worldwide in the oil and gas industries (see Figs 26 and 27).

3.1.7. Containers and packages

In the foreword, it was stated that when DU is used for radiation shielding in sealed source equipment, containers and packages, this publication will refer to these items collectively as 'devices'. However, for clarity, containers and packages are explicitly discussed in this subsection.

Currently, DU alloys are used for gamma ray shielding in containers designed for the storage, transport and disposal of high level waste (HLW) or spent nuclear fuel (SNF), as well as for the transport of high activity, gamma emitting sealed radiation sources [30] (see Figs 28 and 29).

Dual use casks, containing an alloy of DU known as the special uranium alloy BZ-23, have been developed in the Russian Federation for the storage and transport of spent fuel [31]. Another innovative use of DU is to use it in its stable oxide form, mixed with concrete, to make a depleted uranium concrete known as DUCRETE. The DUCRETE provides radiation shielding for dry storage casks used for the storage of spent fuel. Relative to conventional concrete casks, the use of DUCRETE reduces both the weight and the size of the cask, as well as reducing the cask shielding thickness from 71 to 20 cm [31] (see Fig. 30).

FIG. 23. Thickness/density gauge for road surfacing and asphalt, vintage model GDF-30 (courtesy of L. Pillette-Cousin, France).

FIG. 24. Gauges shielded with DU.

FIG. 25. Gauges shielded with DU (courtesy of K. Wee Siang, Nuclear Malaysia).

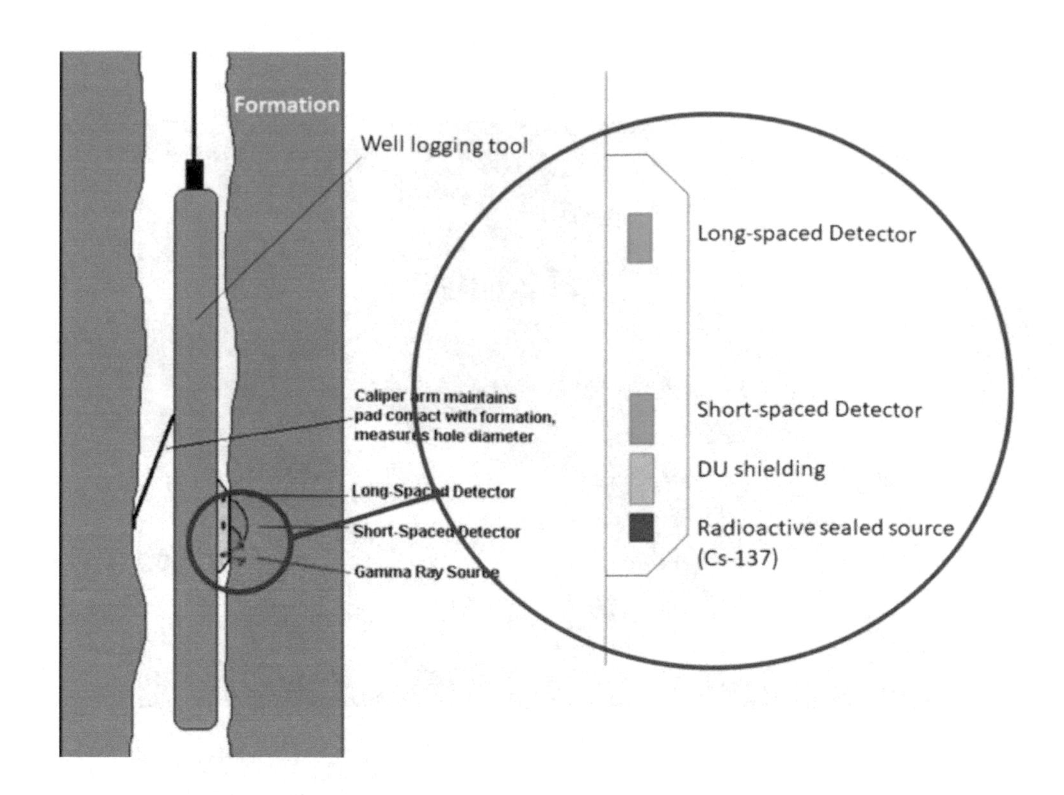

FIG. 26. The principle of a density gamma diagraphy tool (courtesy of L. Pillette-Cousin, France).

FIG. 27. DU blocks used as shielding in well logging tools (courtesy of L. Pillette-Cousin, France).

FIG. 28. High activity gamma source container in a transport configuration (courtesy of Reviss Services).

FIG. 29. High activity gamma source container enclosed in a cage to prevent contact with its hot surfaces (courtesy of Reviss Services).

FIG. 30. Example of a DUCRETE container (courtesy of US Department of Energy).

3.1.8. Linear accelerators

Medical linear accelerators (hereafter referred to as linacs) have been used since the 1970s for cancer radiation therapy. Linacs generate and customize high energy X rays or electron beams to conform to a tumour's shape and destroy cancer cells while sparing the surrounding normal tissue.

Linacs accelerate electrons using a tuned-cavity waveguide, in which a power source creates a standing wave. Some linacs have short, vertically mounted waveguides, whereas higher energy machines tend to have a horizontal, longer waveguide and a bending magnet to turn the beam vertically towards the patient. Linacs use monoenergetic electron beams of between 4 and 25 MeV, giving an X ray output with a spectrum of energies up to and including the electron energy when the electrons are directed to a high density conversion target (such as tungsten). The linac can work in two modes: the photon mode, after conversion of electron pulses into X ray pulses, and the electron mode.

The linac treatment head has a series of collimating, filtering and shielding devices that are used to shape and control the beam and the absorbed dose delivered to the patient. Usually, at the exit of the waveguide generating the accelerated electron pulses, there is a shield called a gun shield back plate. In certain older linac heads, gun shield back plates were made of DU.

A first (primary) collimator is responsible for eliminating the wide-angle radiation generated at the top of the teletherapy head. In some linac heads, parts of the primary collimator have been made of DU.

In the photon mode, the radiation generated by the impact of electrons on the target directs strongly forward, because of the high speed of the incident electrons. This means that the intensity is greater on the axis than on the sides, and therefore any administered dose would not be homogeneous on the treated surface. The role of the flattening filter is to make the X ray flux density more homogeneous. Its inner edge has the shape of a truncated cone pointing towards the impact zone on the target in photon mode. The opening is fixed, which means that it defines the maximum size of the target area at the patient level. In certain older linac heads, flattening filters have been made of DU.

Whereas the primary collimator is fixed and defines the maximum size of the target area, the secondary collimator offers an adjustable opening and serves to narrow the beam over the region to be treated. The secondary collimator is made of two pairs of superimposed jaws that control the opening in one of the two dimensions of the plane perpendicular to the beam. In certain linac heads, parts of the secondary collimator have been made of DU.

Varian low- and single-photon energy linac machines manufactured from the 1970s to the early 1990s often contain DU shielding, including the following models:

— Varian Clinac 4 linacs (manufactured from 1972 to 1990);
— Varian Clinac 4/80 linacs;
— Varian Clinac 4/100 linacs;
— Varian Clinac 6X linacs (manufactured from 1975 to 1978);
— Varian Clinac 6/100 linacs (manufactured from 1979 to 1989);
— Varian 600C models up to serial no. 172 (manufactured from 1989 to 1993).

The number of DU components and the DU quantity by weight varies per machine, but typical amounts found in the Varian Clinac 6/100 would be 44 kg in the primary collimator, 136 kg in the secondary collimator, 8 kg in the faceplate shield and 18 kg in each of the three gun shield discs. Thus, the total DU weight per Varian Clinac 6/100 head is 242 kg. Since the early 1990s, however, DU has been replaced by tungsten.

4. HOW TO IDENTIFY DEVICES CONTAINING DU

This section is intended to help readers identify DU-containing devices using various tools. Online sources of information include the following:

— The International Catalogue of Sealed Radioactive Sources and Devices[1] is a useful tool for identifying unknown devices and containers.
— The European Atomic Energy Community (Euratom) has a database for tracking DU in European Union (EU) countries, including DU used as shielding [32].
— The IAEA has developed and implemented a publicly available database of devices containing DU as shielding for SRSs (hosted at the Disused Sealed Radioactive Sources Network, or DSRSNet) [33] (see Fig. 31). The database provides information on the most typical of these devices to facilitate their identification and proper management.

4.1. DEVICE LABELLING INFORMATION

Often, but not always, a device may be stamped, embossed or labelled with a statement or warning that the device contains DU (may state 'URANIUM'). See Figs 32–34).

4.2. DEVICE DOCUMENTATION

If available, the manufacturer's documentation may indicate that the device or its components contain DU. In addition, the manufacturer could be contacted to get more information about a device that may contain DU. Information on transportation packages that contain DU is included in Ref. [34].

[1] See: https://www.iaea.org/resources/databases/international-catalogue-of-sealed-radioactive-sources-and-devices

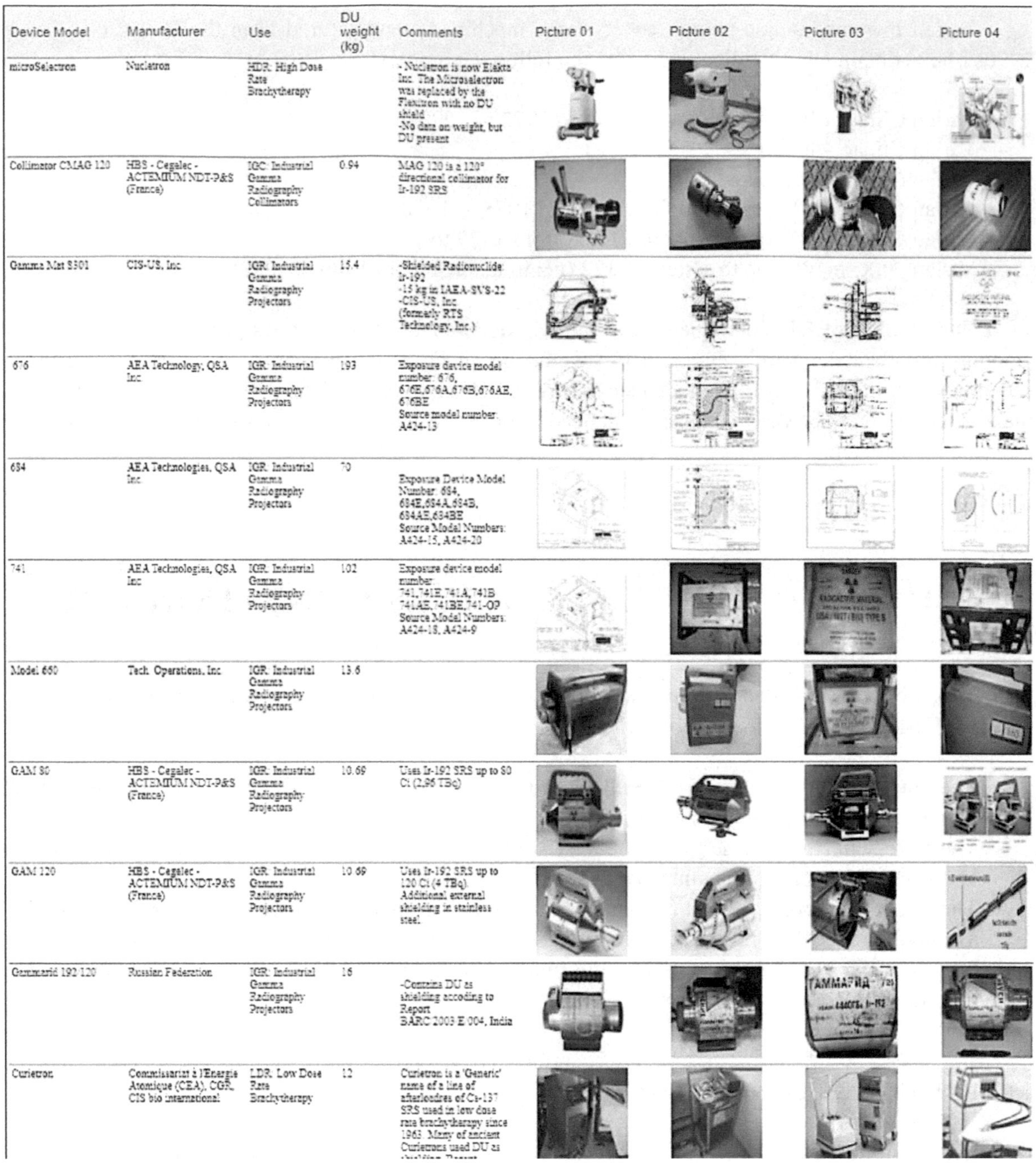

Device Model	Manufacturer	Use	DU weight (kg)	Comments	Picture 01	Picture 02	Picture 03	Picture 04
microSelectron	Nucletron	HDR: High Dose Rate Brachytherapy		- Nucletron is now Elekta Inc. The Microselectron was replaced by the Flexitron with no DU shield -No data on weight, but DU present				
Collimator CMAG 120	HBS - Cegalec - ACTEMIUM NDT-P&S (France)	IGC: Industrial Gamma Radiography Collimators	0.94	MAG 120 is a 120° directional collimator for Ir-192 SRS				
Gamma Mat S301	CIS-US, Inc	IGR: Industrial Gamma Radiography Projectors	15.4	-Shielded Radionuclide: Ir-192 -15 kg in IAEA-SVS-22 -CIS-US, Inc (formerly RTS Technology, Inc.)				
676	AEA Technology, QSA Inc	IGR: Industrial Gamma Radiography Projectors	193	Exposure device model number: 676, 676E,676A,676B,676AE, 676BE Source model number: A424-13				
684	AEA Technologies, QSA Inc	IGR: Industrial Gamma Radiography Projectors	70	Exposure Device Model Number: 684, 684E,684A,684B, 684AE,684BE Source Modal Numbers: A424-15, A424-20				
741	AEA Technologies, QSA Inc	IGR: Industrial Gamma Radiography Projectors	102	Exposure device model number: 741,741E,741A,741B 741AE,741BE,741-OP Source Modal Numbers: A424-18, A424-9				
Model 660	Tech. Operations, Inc	IGR: Industrial Gamma Radiography Projectors	13.6					
GAM 80	HBS - Cegalec - ACTEMIUM NDT-P&S (France)	IGR: Industrial Gamma Radiography Projectors	10.69	Uses Ir-192 SRS up to 80 Ci (2.96 TBq)				
GAM 120	HBS - Cegalec - ACTEMIUM NDT-P&S (France)	IGR: Industrial Gamma Radiography Projectors	10.69	Uses Ir-192 SRS up to 120 Ci (4 TBq). Additional external shielding in stainless steel.				
Gammarid 192/120	Russian Federation	IGR: Industrial Gamma Radiography Projectors	16	-Contains DU as shielding accoding to Report BARC/2003/E/004, India				
Curietron	Commissariat à l'Energie Atomique (CEA), CGR, CIS bio international	LDR: Low Dose Rate Brachytherapy	12	Curietron is a 'Generic' name of a line of afterloadres of Cs-137 SRS used in low dose rate brachytherapy since 1963. Many of ancient Curietrons used DU as				

FIG. 31. Screenshot of the IAEA database of devices containing DU as a radiation shielding material.

FIG. 32. Label on a device containing DU with the words 'This device contains approximately 26 pounds of depleted uranium'.

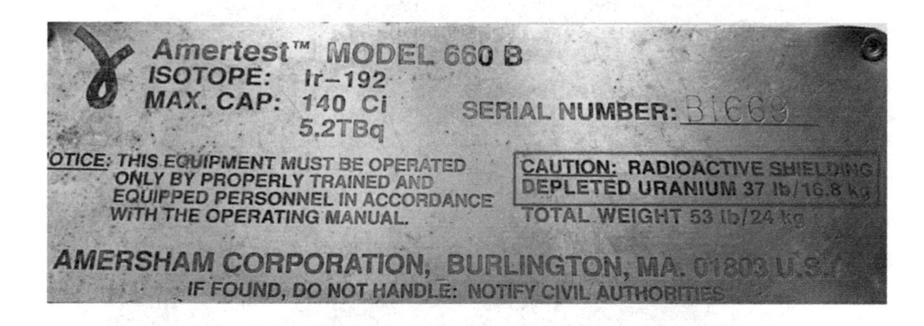

FIG. 33. Label on a radiography camera containing DU with the words 'Caution: radioactive shielding depleted uranium 37 lb./16.8 kg'.

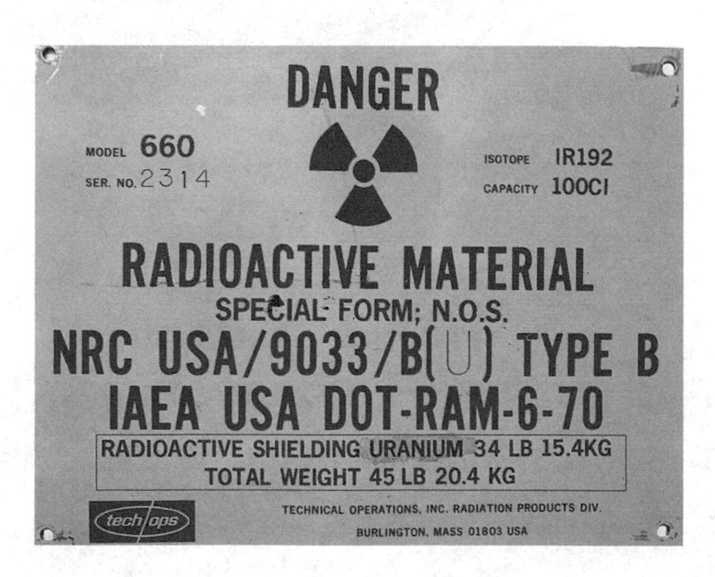

FIG. 34. Label on a device containing DU with the words 'Radioactive shielding uranium 34 lb. 15.4 kg'.

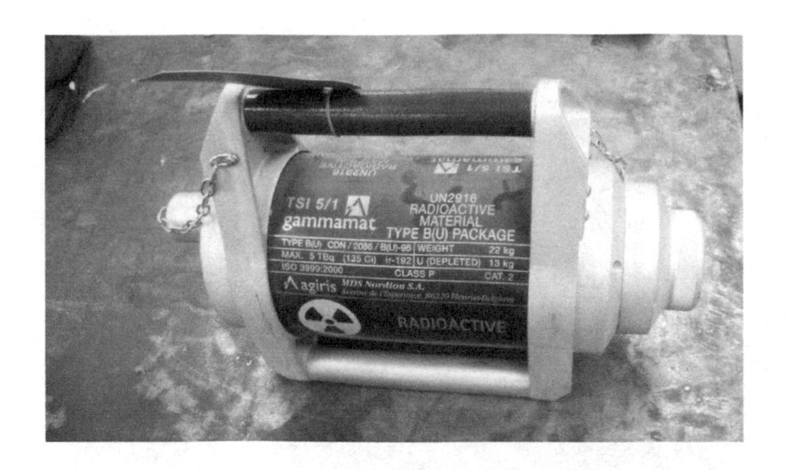

FIG. 35. Label on a radiography projector with the words 'U (depleted) 13 kg'.

FIG. 36. Label on a radiography projector Sentinel 660, with the label words, 'Radioactive shielding depleted uranium 16.8 kg'.

FIG. 37. Parts of a DU collimator with the embossed word 'uranium'.

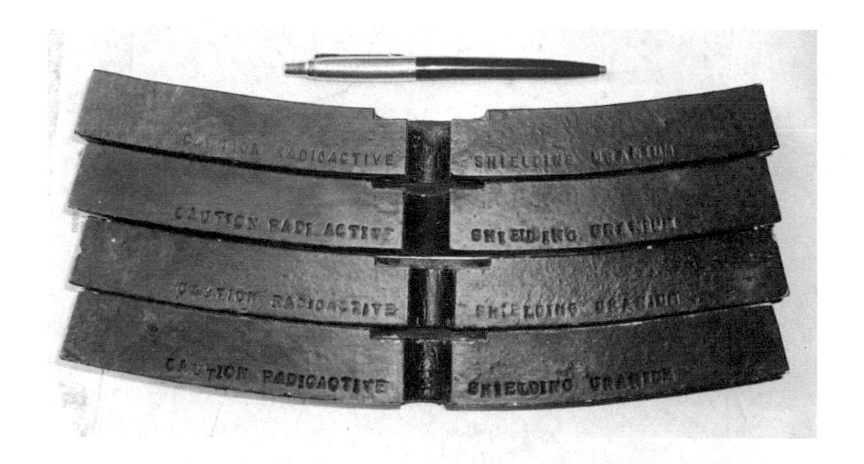

FIG. 38. *Parts of a DU collimator with the embossed word 'uranium'.*

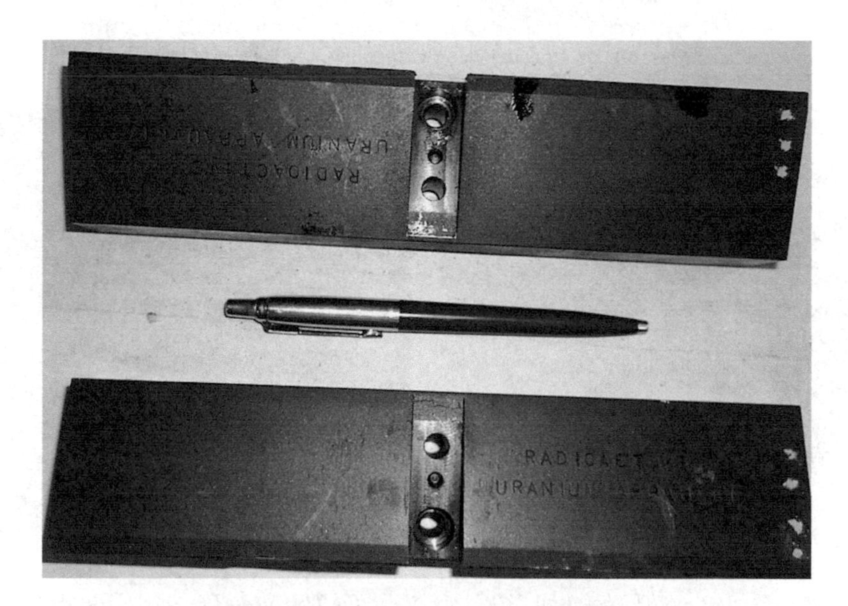

FIG. 39. *Parts of a DU collimator marked with the word 'uranium'.*

FIG. 40. *Uranium shielding markings on a teletherapy head with the word 'uranium'.*

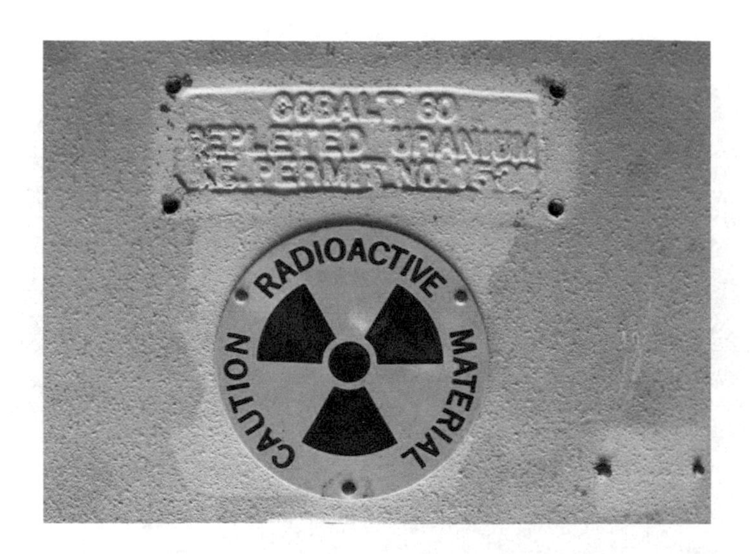

FIG. 41. Uranium shielding markings on a teletherapy head with the words 'depleted uranium' embossed on the container.

FIG. 42. (Left) Teletherapy unit model Theratron 780C (applies to the Theratron Phoenix) with (right) tag indicating DU presence.

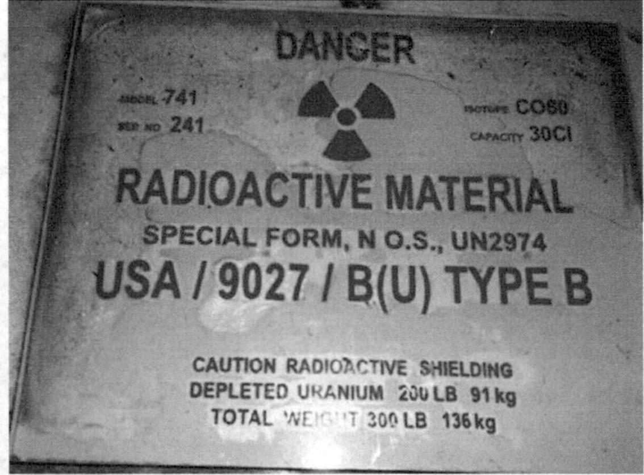

FIG. 43. (Left) Model 741 mobile industrial radiography exposure device; (right) 91 kg of DU is indicated on the tag.

FIG. 44. (Left) A transport cask with (right) the embossed words '9 kg of depleted uranium' (courtesy of Global Medical Isotope Systems).

4.3. DEVICE EXAMINATION TO DETERMINE IF IT CONTAINS DU

Any empty device that contained radioactive source(s) or was a linac needs to be examined to see if the device or its components contain DU. Examples of such devices are shown in Section 4.1 of this publication. Any device that contains DU will have a radiation dose rate reading of ≥2 μSv/h above background level when a survey meter's probe is held close to the DU with little or no lead or steel shielding between the probe and the DU. An example of the radiation dose rate near a teletherapy device's trimmer bar is shown in Fig. 45. This measurement of the low radiation dose rate has to be performed after the source(s) have been removed from the device. Detecting DU in linac shielding or collimators can be done if there is no significant high energy photon activation of the accelerator's components.

Gamma spectrometry can be used to confirm whether the material is made of DU or whether there is radioactive source contamination or activation in an accelerator (see Figs 46–50).

Radiation measurements have to be performed by qualified personnel.

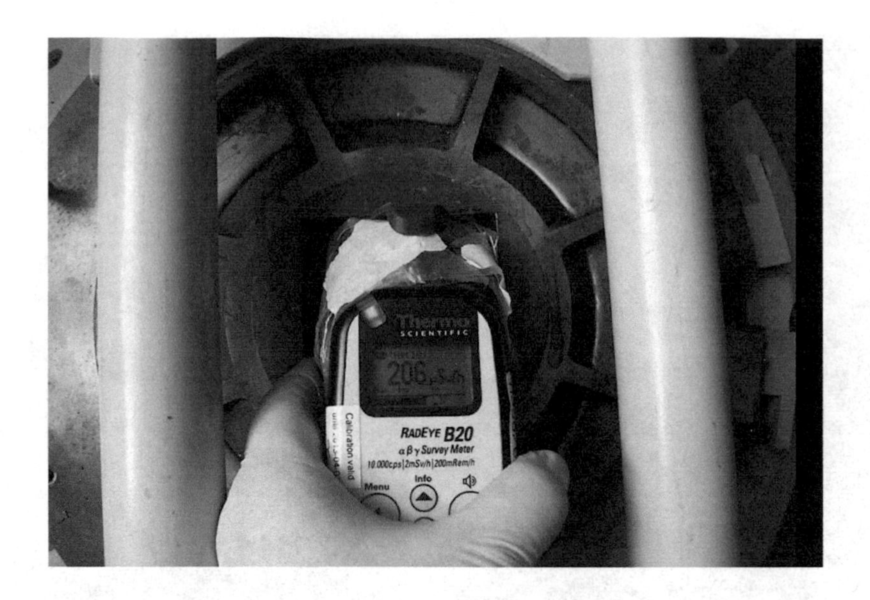

FIG. 45. Measuring the radiation dose rate near the DU collimators of a Model C9 teletherapy unit.

FIG. 46. Part of a DU collimator identified as uranium by gamma spectrometry.

FIG. 47. A DU collimator identified as uranium by gamma spectrometry.

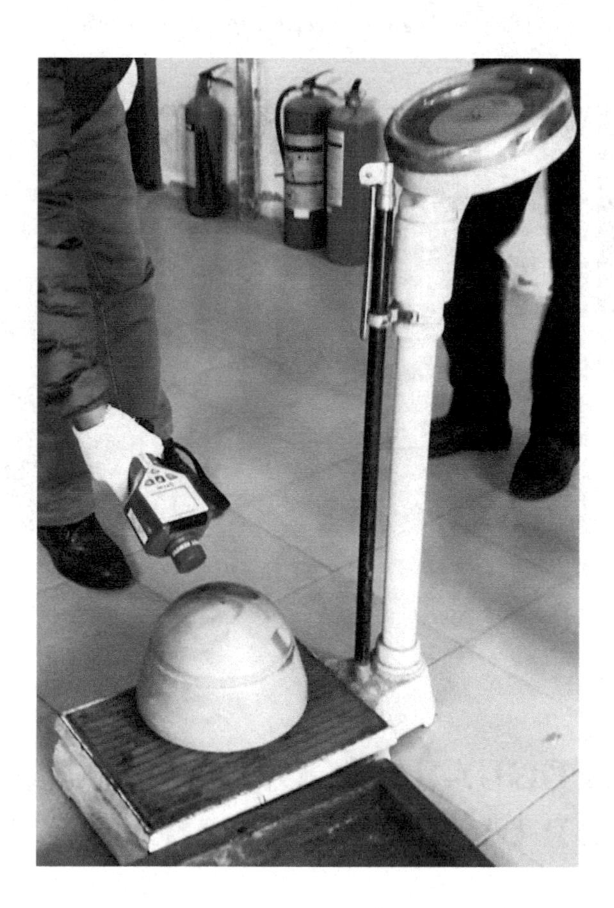

FIG. 48. A DU block/shielding identified by gamma spectrometry.

FIG. 49. Example of identification of DU shielding with a gamma spectrometer when the information on the tag is not available or visible.

FIG. 50. Level gauge from China containing DU.

5. SAFEGUARDS CONSIDERATIONS FOR DU SHIELDED DEVICES

States conclude safeguards agreements with the IAEA in order to fulfil their non-proliferation commitments in connection with the Treaty on the Non-Proliferation of Nuclear Weapons (NPT) [35]. The IAEA applies safeguards pursuant to three types of safeguards agreements: comprehensive safeguards agreements (CSAs) and additional protocols concluded in connection with the NPT and/or treaties establishing nuclear weapon free zones (for ~170 States); item-specific safeguards agreements (for three States); and voluntary offer agreements (for the five nuclear weapon states party to the NPT).

5.1. OBLIGATIONS AND RESPONSIBILITIES

Each non-nuclear weapon state party to the NPT is required to conclude a CSA with the IAEA. A model agreement based on INFCIRC/153 (Corrected) [35] is published as GOV/INF/276, annex I [36]. The vast majority of States have concluded a CSA with the IAEA. The rest of this section focuses on the safeguards obligations related to DU metal used as radiation shielding in such States.

Under a CSA, the State undertakes to accept IAEA safeguards in accordance with the terms of the safeguards agreement, on all source or special fissionable material in all peaceful nuclear activities within the territory of the State, under its jurisdiction or carried out under its control anywhere. The IAEA has the corresponding right and obligation to ensure that such safeguards are applied to all such material, for the exclusive purpose of verifying that it is not diverted to nuclear weapons or other nuclear explosive devices.

Nuclear materials subject to safeguards are defined in article XX of the IAEA's Statute:

— Special fissionable materials are 239Pu, 233U and uranium enriched in isotopes 235 or 233;
— Source material includes natural uranium, DU and thorium in any physical or chemical form.

DU is included in the definition of 'source material' and thus is considered nuclear material subject to safeguards. As such, it needs to be declared.

States with a CSA in force are required to provide information to the IAEA about inventories and flows of all nuclear material (including DU) and to facilitate access by the IAEA to conduct inspections at facilities and locations where such material is present. Therefore, when managing DU, it is important to ensure the relevant safeguards obligations under a CSA are met. This section describes these obligations in general terms. For a more complete explanation of safeguards obligations, detailed guidance is provided in the Guidance for States Implementing Comprehensive Safeguards Agreements and Additional Protocols found on the Assistance for States web page[2] [37].

Nuclear material, including DU, continues to be subject to safeguards until such time as it is determined by the IAEA that safeguards can be terminated on such material. INFCIRC/153 (Corr.) states in paragraph 11 that

> "... safeguards shall terminate on nuclear material subject to safeguards thereunder upon determination by the Agency that it has been consumed, or has been diluted in such a way that it is no longer usable for any nuclear activity relevant from the point of view of safeguards, or has become practicably irrecoverable."

As such, DU in the form of metal used as shielding will not meet the requirements for termination of safeguards.

States with a CSA have the right to request that nuclear material, including DU (that was previously reported by the State to the IAEA under safeguards) be exempted from safeguards. Exemption may be requested for such nuclear material on the grounds that it is either a small quantity (less than one effective kilogram) or that it is used for a non-nuclear purpose (such as a counterweight in a crane or shielding in a container). If the IAEA grants the exemption, the State is not required to submit accounting reports

[2] Many States have concluded an 'Additional Protocol' to their safeguards agreements. Such States undertake to provide additional information and access to the IAEA, to strengthen the effectiveness and efficiency of safeguards. States with very limited quantities of nuclear material may conclude a 'small quantities protocol' to their CSA, which holds in abeyance, or suspends, some of the safeguards procedures in Part II of the CSA [16].

in respect of this nuclear material, and the Agency no longer routinely verifies it. However, the nuclear material remains subject to safeguards, and pursuant to an additional protocol, the IAEA retains rights of access to certain locations containing exempted material.

If exempted nuclear material is to be processed or stored together with non-exempted nuclear material or if it is to be exported outside of the State, the State authority that is responsible for safeguards must arrange in advance for the reapplication of safeguards to that nuclear material. In such cases, the State authority must send a letter to the IAEA requesting the de-exemption of the relevant items. (The term 'exempted' in the field of safeguards is not the same as the one in the regulatory domain, such as commonly used in the context of radiation safety). Nuclear material exempted from safeguards should always remain subject to the national regulatory control of the State authority. Appendix II provides a case study regarding the control and protection of DU shielding in France through the declaration regime.

As noted above, prior to shipping exempted DU out of a State, the State authority must first request de-exemption from IAEA safeguards. Following de-exemption, the export of DU is reported to the IAEA as an 'inventory change report'. After export/shipment, the DU becomes part of the safeguards inventory of the importing State. The importing State may subsequently, at its discretion, request that the DU be exempted from safeguards on the grounds set out above.

5.2. STEPS TO TAKE WHEN DEALING WITH DU SHIELDS

When a DU shield that is not in a national inventory is found or identified, the first action should always be to document the material and secure it to avoid accidental theft or mishandling (e.g. by placing it in a locked room). Once secured, the State or regional safeguards authority should be contacted to verify the status of the material. It is possible that the material is in full compliance if the shield has a transit exemption from safeguards (i.e. if it still belongs to the originating State). If the material does belong to the State in which it was found, there is a process to add the material as an accidental gain per guidance in the Nuclear Material Accounting Handbook, IAEA Services Series No. 15 [38]. This process is typically pursued under the guidance of the State's safeguards authority in consultation with the IAEA and is generally seen as a normal component of a safeguards programme. In all cases, however, the main recommendation is to keep the lines of communication open between the location of the item, the State safeguards authority and the IAEA. This communication will keep everyone informed throughout the process and result in a smoother application of the appropriate safeguards for the material. In every case, the appropriate steps to remember when dealing with material with an unknown safeguards burden is to Secure, Document, Notify and Declare.

6. SAFETY AND SECURITY CONSIDERATIONS FOR DU SHIELDED DEVICES

The IAEA has issued Safety Standards and security guidance publications emphasizing the necessity of national programmes to ensure the safety and security of DSRSs [10, 39, 40], including those containing DU as shielding material.

To ensure the effective management and control of radioactive waste, the government shall ensure that a national policy and a strategy for radioactive waste management are established [10], including a policy and strategy on DSRS management.

Precautions need to be taken when handling, using, transporting and storing DU material. These are usually the same as with other solid radioactive waste or material. Radiation protection procedures have to be implemented for workers handling, transporting, recycling and storing DU shielding.

The facility where a device is handled or stored has to be equipped with appropriate capabilities for device handling, including lifting/hoisting tools, adequate lighting and power sources, and remote and handheld radiation and contamination monitoring instrumentation. The complexity and sophistication of handling capabilities are directly related to the potential hazards of the devices to be handled.

When a device is handled, some precautions need to be taken to prevent mechanical damage to the device (e.g. protecting it from being dropped from a height or struck by a handling machine or forklift) and to ensure the functionality of the safety systems of the source holder (e.g. checking the locking mechanism of the sources is intact).

Some specific industrial safety practices include:

— Handling heavy DU devices only under the supervision of a person qualified in the safe movement of heavy items. The appropriate safety equipment needs to be used (e.g. steel toed boots, back harnesses, hand trucks).
— Maintaining the integrity of the device, e.g. do not cut, grind or abrade the DU (to eliminate pyrophoric hazard).
— Using proper firefighting techniques for DU fires. Water spray, CO_2 and halon extinguishers are ineffective at fighting uranium fires [41].

6.1. RADIOLOGICAL SAFETY

The requirements of GSR Part 3 [23] and national regulations have to be implemented. The application of these standards by EU Member States is ensured by the implementation of binding Euratom legislation [42]. The 'as low as reasonably achievable' (ALARA) principle needs to be applied [43].

It is also necessary to minimize the dose delivered to workers by the comprehensive use of time, distance and shielding:

— Time: minimize the amount of time spent near a radioactive source;
— Distance: maximize the distance from a radioactive source as much as possible;
— Shielding: place shielding between the radioactive source and the worker.

The implementation of the following GSR Part 3 [23] requirements results in radiation protection measures applicable in the case of DU devices:

— Requirement 24 of GSR Part 3 states:

"Employers, registrants and licensees shall establish and maintain organizational, procedural and technical arrangements for the designation of controlled areas and supervised areas, for local rules and for monitoring of the workplace, in a radiation protection programme for occupational exposure."

— Para. 3.90 of GSR Part 3 states:

"Registrants and licensees:

(a) Shall delineate controlled areas by physical means or, where this is not reasonably practicable, by some other suitable means.

…

(c) Shall display the symbol recommended by the International Organization for Standardization and shall display instructions at access points to and at appropriate locations within controlled areas."

- Requirements 24 and 30 of GSR Part 3 give provisions for cases with a risk of spread of contamination of workers or the public.
- Para. 3.43 of GSR Part 3 states:

"If the safety assessment indicates that there is a reasonable likelihood of an emergency affecting either workers or members of the public, the registrant or licensee shall prepare an emergency plan for the protection of people and the environment."

Devices will be stored in a manner that minimizes potential contamination of the work area. Devices may be stored in their original package assemblies, since these provide suitable radiation protection (this type of container will retain and protect the contents during normal transportation activities). Drums, steel boxes (avoid wood and cardboard because of contamination and biological degradation concerns) and similar containers might be used to store DU, which has the potential to release DU oxide contamination. Control of contamination is key to protecting the workforce from internal radiation hazards. Practising good hygiene and housekeeping habits mitigates internal radiation hazards.

Contain any devices with loose DU oxides (contamination) such as by placing a complete device in a plastic bag or another airtight container. Any oxidation layer should not be removed because of the contamination hazard, except by properly trained and skilled personnel in contamination controlled environments such as gloveboxes.

Areas with known radiological contamination have to be clearly marked, so workers can take actions to minimize personal contamination. The use of personal protective equipment such as protective suit, lab coats, gloves and foot coverings minimizes the possibility of contamination.

6.2. SAFETY AND SECURITY FOR THE STORAGE OF DU SHIELDED DEVICES

The safe and secure storage of DU is important to ensure that nuclear material is not diverted from peaceful use, in accordance with NPT commitments, as well as to protect workers and the public from the danger of ionizing radiation. To accomplish these goals:

- The material will be stored in a secure area with controlled access. The level of security is determined by a graded approach. Based upon a risk and threat assessment, the requirements for access control are determined. Lock and key condition with effective key control is often enough to maintain a secure storage facility.
- Conditions to prevent corrosion of the DU shields need to be implemented. For instance, dry conditions and a closed storage area are recommended for storage of DU. Temperature control may require consideration.
- Contamination risk has to be minimized, e.g. by keeping devices in closed and airtight drums and boxes, and making routine contamination checks.
- A security plan should be implemented that addresses administrative and technical security measures to protect against unauthorized removal or malicious actions.
- Fire suppression equipment compatible with uranium fires may be available, and/or firefighters might be informed of the special considerations for DU.
- Tamper evidence security seals could be used on devices and packages containing DU or to seal storage cabinets and rooms where DU is stored. If the seal is intact, it means that the content is

still the same as when the seal was installed. This may simplify periodic and annual verification of inventories.
- A contingency plan to address unauthorized access has to be implemented.
- Any devices or packages marked as empty (SRS removed) will be checked to ensure that they are indeed empty. A verification indicator will be attached (e.g. verified empty on date = DD.MM. YYYY).
- If there is a reasonable probability of an emergency related to DU storage, a licensee needs to have a DU specific emergency plan for the protection of people and the environment.
- Existing historical documentation of the empty packaging and articles has to be collected, processed and maintained.

IAEA Nuclear Security Series No. 11-G (Rev. 1), Security of Radioactive Material in Use and Storage and of Associated Facilities [40], No. 13, Nuclear Security Recommendations on Physical Protection of Nuclear Material and Nuclear Facilities [44], No. 14, Nuclear Security Recommendations on Radioactive Material and Associated Facilities [45] and No. 27-G, Physical Protection of Nuclear Material and Nuclear Facilities [46] provide guidance on the protection of nuclear and other radioactive material. According to these publications, the DU used as shielding is a form of radioactive material that would present very low potential radiological consequences if it was subjected to unauthorized removal or sabotage; thus, DU does not represent a substantial security concern. A security level should not need to be assigned to DU, but it needs to be protected through prudent management practices during its use and storage.

National experiences in implementing safe and secure measures for managing DU are described in the annexes. A national experience (Hungary) specifically addressing the physical protection level requirements for the storage of DU is described in Appendix III.

6.3. SAFETY AND SECURITY FOR THE TRANSPORT OF DU SHIELDED DEVICES

The principal IAEA Safety Standards applicable to the transport of radioactive material are Specific Safety Requirements (SSRs), Specific Safety Guides (SSGs) and Safety Guides (TS-Gs), for instance:

- IAEA Safety Standards Series No. SSR-6 (Rev.1), Regulations for the Safe Transport of Radioactive Material (2018 Edition) [29];
- IAEA Safety Standards Series No. SSG-26, Advisory Material for the IAEA Regulations for the Safe Transport of Radioactive Material (2012 Edition) [47];
- IAEA Safety Standards Series No. SSG-33 (Rev. 1), Schedules of Provisions of the IAEA Regulations for the Safe Transport of Radioactive Material (2018 Edition) [48];
- IAEA Safety Standards Series No. SSG-65, Preparedness and Response for a Nuclear or Radiological Emergency Involving the Transport of Radioactive Material [49];
- IAEA Safety Standards Series No. TS-G-1.3 Radiation Protection Programmes for the Transport of Radioactive Material [50];
- IAEA Safety Standards Series No. TS-G-1.4, The Management System for the Safe Transport of Radioactive Material [51];
- IAEA Safety Standards Series No. TS-G-1.5, Compliance Assurance for the Safe Transport of Radioactive Material [52].

IAEA Nuclear Security Series No. 9-G (Rev. 1), Security of Radioactive Material in Transport [53], and No. 26-G, Security of Nuclear Material in Transport [54] provide guidance on the protection of nuclear and other radioactive material during transportation.

The primary goal for the transport of disused DU devices is to ensure safety; that is, to protect people, property and the environment from the effects of radiation (para. 104 of SSR-6, (Rev. 1)) [29]. This goal is mainly achieved by:

— Containment of DU;
— Control of external dose rate;
— Prevention of damage caused by fire.

For Member States that require assistance with the transport of disused DU devices, assistance can be sought from the IAEA, as well as Member States with experience transporting disused DU devices.

See Figs 51–53 for examples of the preparation of DU devices for transport.

FIG. 51. Preparation of a teletherapy head containing DU and ^{60}Co source for international transport.

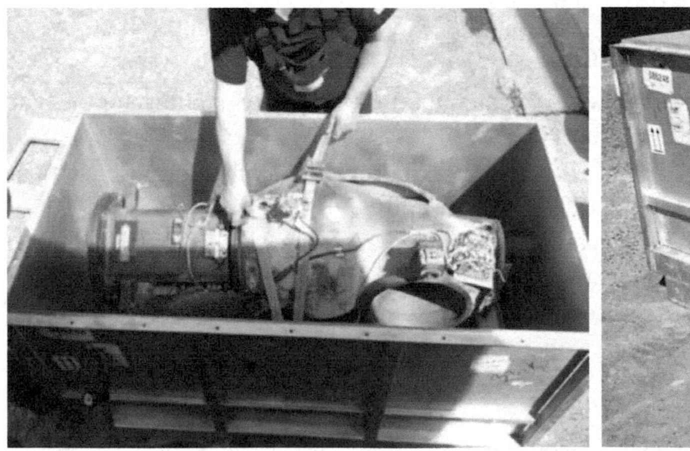

FIG. 52. Preparation of a teletherapy head containing DU as shielding material and without SRSs for transport.

FIG. 53. Removal of collimators with DU from a teletherapy head and preparation for transport.

Provided that compliance with the SSR-6 (Rev. 1) transport regulations is ensured in all cases, DU could be classified for transport as one of the following [29]:

— UN 2908, Radioactive material, excepted package — empty packaging;
— UN 2909, Radioactive material, excepted package — articles manufactured from natural uranium or depleted uranium or natural thorium;
— UN 2911, Radioactive material, excepted package — instruments or articles;
— UN 2912, Radioactive material, low specific activity (LSA-I), used for cases if the DU is not covered by an inactive sheath made of metal or some other substantial material or if the dose rate on the surface of the package is >5 μSv/h, in which case a type IP-1 package is required for transportation.

Transport requirements can be complex, and a variety of options require consideration. For instance, disused DU devices can be transported under UN 2909 if the dose rate from the package is below 5 μSv/h on contact with the package. If the dose rate is more than 5 μSv/h on contact, it needs to be transported under UN 2912. Another consideration is the size of the package. If a suitably sized container is not available for an item such as a teletherapy head, transporting a device under special arrangement modality could be proposed, which is an exceptional case accepted by IAEA transport regulations.

A disused gamma radiography projector, which itself may be considered a package, may be transported under UN 2908, as an empty package, if the dose rate at contact is less than 5 μSv/h. If the dose rate at the surface is more than 5 μSv/h, the projector may be located inside an excepted package to reduce the dose rate. Now, the projector is not transported as an empty package but as an excepted package under UN 2909. The dose rate at the surface of the package containing the projector has to be verified for compliance with the limit of 5 μSv/h.

Refer to the Hungarian national report in Annex VIII for another example. See Appendix III for an example of practices for DU transport (Hungary).

7. MANAGEMENT OPTIONS FOR DU

7.1. GENERAL INFORMATION

As with any hazardous material (see Section 2), it is important that appropriate and safe management options are considered and developed, based on the material's physical and chemical characteristics, total inventory, applicable regulations, available technology and funding.

The variation of DU inventories in Member States makes it difficult to develop or recommend a common management plan for DU from disused devices for any given Member State.

In general, Member States with significant DU inventories have yet to fully implement policies or strategies with regard to long term DU management. Member States with smaller quantities of DU sometimes adopt a wait-and-see position with regard to the long term management of DU. However, a variety of short term issues need to be promptly addressed regarding DU from shielded devices and containers, as discussed next.

A management system should be applied to the management of devices (equipment, containers and packages) containing DU material [55].

7.2. INVENTORY INFORMATION

Member States need to establish and maintain, to the extent practicable, an inventory of devices containing DU in their jurisdictions. This can be implemented as part of an existing inventory system in Member States. The operator is responsible for establishing and maintaining a detailed record-keeping system for all stages of radioactive waste management [56]. Data should be regularly reviewed and verified. Some examples are provided in the national reports in the annexes.

Information, for each device in the inventory, includes as a minimum:

- Current location;
- Type of device or container, e.g. blood irradiator, gamma radiography projector, etc.;
- Manufacturer, model and serial number;
- Date of possession;
- Device documentation (dimensions, drawings, certificates, authorizations, licences);
- Total weight of the device and DU;
- Overpack information (type, dimensions);
- Photographs;
- Contamination survey results;
- Dose rate measurement results;
- Information about the user/former user, e.g. organization, location, contact;
- All relevant information about SRSs that are still within devices (including leak tests if applicable).

7.3. MANAGEMENT OPTIONS FOR DEVICES CONTAINING DU AS SHIELDING MATERIAL

Management options for disused devices containing DU as a radiation shielding material include return to supplier (manufacturer/distributor), reuse or recycling, long term storage and disposal.

Each Member State has to follow its own regulations related to DU. If specific regulations do not exist, management procedures need to be coordinated with the regulatory body.

7.3.1. Return to manufacturer/supplier

The export of DU for recycling, reuse or disposal to another country, regardless of quantity, will require reporting to the IAEA in accordance with safeguards requirements and the Member State's regulations [35]. DU is shipped between countries for recycling and/or disposal in accordance with the IAEA transport regulations, as well as national, bilateral or international agreements.

Ideally, each Member State may wish to weigh the benefits of establishing and implementing a 'return to manufacturer/supplier' policy for DU shielded devices, along with disused sources, as part of the initial device purchase agreement. All new contracts for the purchase of devices and SRSs might include a clause that calls for the return of these devices, along with the disused source, to the manufacturer.

The return of devices to a manufacturer or supplier, with or without the disused sources, is the preferred management option. If not agreed upon initially, the return could still be negotiated prior to considering other management options.

The return option has to take into consideration the costs involved in the packaging and transportation of relatively heavy devices (see Section 6.3), in accordance with applicable IAEA and/or national packaging and transportation regulations [29]. These costs can be significant. In some cases, this has served as a deterrent and hindered the return of such devices to manufacturers. The organization accepting the return may sometimes assume responsibility for decommissioning these devices. This may incur additional costs.

Some examples of manufacturers or suppliers that have a 'return to manufacturer/supplier' agreement include:

— MDS Nordion in Canada [57];
— UJP PRAHA in the Czech Republic[3];
— QSA Global in the United States of America (USA)[4].

7.3.2. Reuse or recycle

Reuse is simply the transfer of devices, without modification, to the same or another manufacturer for their original purpose. Recycling involves some degree of modification before the DU can be used for some other purpose(s).

It is important to note that most Member States have relatively small inventories of DU shielded devices. Economic factors, besides the absence of a management policy, are a major driving force for considering the reuse or recycle options over other management options. For relatively small inventories, it may be economical and cost effective to recycle DU in a licensed recycling/processing facility, even in another Member State. With this option, one has to take into consideration the associated packaging and transportation regulations and costs involved in shipping the DU to the recycling/processing facility.

The selection of the reuse or recycling options can be justified on the basis of a cost–benefit analysis, which takes into account factors such as:

— Purchasing replacement equipment with a return of the old device included;
— The size of the existing DU inventory (which affects the economics);
— Availability of recycling facilities in the user's Member State or in another Member State that will accept the exported devices;
— Other incurred costs, such as decommissioning, packaging and transportation.

---

[3] See: http://www.iupesm2018.org/resource/iupesm/admin/filegroupfile/SV-1%20UJP%20Prague_1494700813168.pdf

[4] See: https://www.qsa-global.com/radioactive-source-disposition

Devices returned to suppliers/manufacturers for reuse are subject to the same radiological and safeguards requirements as the originally installed equipment. Refer to Section 5 regarding safeguards requirements.

Some companies licensed for reuse and or recycling include:

- Gamma-Service Recycling in Germany[5];
- UJP PRAHA in the Czech Republic;
- Manufacturing Sciences Corporation in the USA[6];
- TOXCO Material Management Center in the USA[7].

7.3.3. Storage

Prior to considering any option to recycle, reuse or dispose of devices, it is important to keep in mind that the DU will need to be stored as part of its life cycle management. Storage is subject to national regulatory requirements.

Specific guidance on regulatory requirements for the design and operation of waste processing and storage facilities is given in IAEA Safety Standards Series No. SSG-45, Predisposal Management of Radioactive Waste from the Use of Radioactive Material in Medicine, Industry, Agriculture, Research and Education [56], IAEA Safety Standards Series No. SSG-40, Predisposal Management of Radioactive Waste from Nuclear Power Plants and Research Reactors [58], and IAEA Safety Standards Series No. WS-G-6.1, Storage of Radioactive Waste [59].

Storage times can vary greatly, depending on the management option selected. For example, for the return option storage times may be relatively short, and for the disposal option storage times may be relatively long. Many Member States already have proper facilities for the storage of DU shields.

Examples of DU storage are provided in national reports in the annexes.

7.3.4. Disposal

Waste disposal means the "emplacement of waste in an appropriate facility without the intention of retrieval" [1]. A disposal facility needs to provide passive safety after its closure, meaning that no active safety or security measures are required to ensure that it maintains its isolation and containment functions [11]. Disposal facilities use a combination of natural and engineered barriers to contain and isolate radionuclides so that they do not cause an unacceptable impact to health and the environment, now or in the future [11]. A disposal facility is sited to isolate the waste from natural or human disturbances and to ensure that the characteristics of the environment will function together with the engineered barriers to provide adequate containment.

Different kinds of waste require varying periods of isolation from people and the environment, and thus they need different disposal solutions. For instance, waste with long lived radionuclides requires longer periods of containment and isolation than other types of waste. The IAEA's waste classification scheme provides a direct link between the waste class and the suitable disposal options [60].

Several near surface disposal facilities are being operated worldwide. A near surface disposal facility is located at, or within a few tens of metres of, the Earth's surface [61]. These facilities offer a safe disposal solution for LLW. However, intermediate level waste (ILW) and high level waste (HLW), along with SNF, cannot be sufficiently isolated and contained in near surface disposal facilities and needs to be disposed of in geological disposal facilities [62]. A geological disposal facility is located underground, usually several hundred metres or more below the Earth's surface, in a stable geological formation to provide long term isolation of radionuclides from the biosphere.

---

[5] See: https://www-ns.iaea.org/downloads/rw/waste-safety/workshops/lisbon2010/tuesday/gamma-service-recycling.pdf

[6] See: https://www.mfgsci.com/materialsacquisition

[7] See: https://www.toxcommc.com/

Different disposal concepts that have been implemented or are being explored for different types of radioactive waste are:

- Trenches: These are facilities similar to conventional landfills for industrial or domestic waste. They are generally implemented in arid regions with low rainfall and deep groundwater. They can offer a disposal solution for very low level waste (VLLW) and LLW.
- Engineered structures for the disposal of LLW: These are facilities built on or just below the surface. Typically, they consist of an array of concrete vaults in which the waste packages are emplaced.
- Underground caverns or silos: Facilities at depths of the order of tens of metres to a few hundred metres can be used as disposal facilities for LLW and ILW. Their detailed design depends on the waste type and geological environment. After waste emplacement operations are completed, the facility is generally backfilled and sealed.
- Geological disposal: The disposal of waste in a stable geological environment at depths of several hundred metres is the only disposal concept that can provide a safe solution for HLW or SNF. These wastes need to be isolated from the biosphere for tens or hundreds of thousands of years. It is not possible to ensure this level of isolation in facilities at the surface or at depths of only a few tens of metres. Designs for geological disposal repositories have been developed for a range of geological environments, such as salt, crystalline rock and clay formations.
- Borehole disposal: Borehole disposal entails the emplacement of waste canisters in one or more boreholes. Concepts for borehole disposal have considered disposal depths ranging from a few tens of metres up to several hundred metres or even kilometres. The diameter may vary from tens of centimetres up to a few metres. After lowering the waste canisters down the borehole, the borehole will be backfilled and sealed.

More comprehensive descriptions of these waste disposal concepts can be found in Refs [63] and [64].

Member States that are operating or planning disposal facilities for waste other than DU may consider co-disposal of DU in these facilities. This will require verifying that the DU complies with the waste acceptance criteria of these facilities.

The approaches to DU disposal applied in different Member States are discussed below. More information can be found in the annexes. These cases are included by way of example, and a disposal solution that is implemented in one Member State may not always be the best option for another one. Differences in the legal and regulatory framework may exclude options applied in other countries, and the preferred disposal route also depends on the DU inventory (e.g. volume and concentration), national policies, available technical options, preferences of the major interested parties, etc.

It is important that the factors affecting the suitability of the disposal options are identified and understood. The disposal of DU needs to be embedded in a wider national policy and strategy on waste management. Guidance on developing a waste management policy and strategy is given in Ref. [65].

The following DU disposal examples may therefore not be applicable to other Member States. They need to be regarded as illustrations of possible technical solutions for the disposal of DU.

Czech Republic

The Radioactive Waste Repository Authority (RAWRA), the State body responsible for radioactive waste disposal in the Czech Republic, has examined the possibility of disposing of DU in the Richard repository. This repository, a former limestone mine at a depth of 30 m to 70 m, has been in operation since 1964 for the disposal of institutional LLW [66].

This disposal route was, however, discarded. DU is subject to IAEA safeguards. RAWRA can therefore dispose of DU only when it is transformed to unrecoverable form and released from safeguards supervision. Disposal in the Richard repository is not considered to be unrecoverable and it therefore does not meet the safeguards requirements.

France

In France, the National Agency for Radioactive Waste Management (Andra) does not accept disused DU shielded devices at its disposal facilities for LLW and ILW (CSFMA) or for VLLW (CIRES) because these devices do not comply with the waste acceptance criteria of these facilities. The facilities are licensed to dispose of waste contaminated with DU in limited quantities but not DU metal blocks. Therefore, Andra is considering a project for a near surface or intermediate depth repository dedicated to long lived low level waste (LL-LLW). This would have the potential to contain tens of thousands of tonnes of radium and thorium bearing waste, uranium conversion residues, etc., and the French inventory of about 100 tonnes of DU devices would be easily disposed of.

According to the French Alternative Energies and Atomic Energy Commission (CEA), which represents the French Government for the management of DU devices, the reference solution would be disposal in a future Andra LL-LLW facility. A preferred solution for the elimination of disused DU shielding would be recycling. It is currently a more expensive solution, and there are uncertainties attached to the cost evolution of recycling, storage and disposal, but from the societal and environmental standpoints, recycling will be better for future generations [67]. See also the French national report in Annex VI.

Germany

In Germany, DU is currently considered as a resource and not as waste [68]. If, in the future, DU were to be classified as waste, it would require geological disposal. A former iron mine, the Konrad mine, is being converted into a disposal facility for radioactive waste with negligible heat generation (LLW and ILW). This facility's licence, however, limits the amount of uranium that can be disposed of and does not allow the disposal of large stocks of DU.

A preliminary safety assessment was performed for the disposal of 35 000 m^3 of DU in a salt formation at Gorleben, a site that was formerly considered as a possible site for the disposal of heat generating waste. The site is no longer being considered in the ongoing siting process for a geological disposal facility.

United Kingdom

The United Kingdom (UK) has also been assessing options for the long term management of DU. Policies for managing large quantities of DU from enrichment plants would undoubtedly accommodate DU from shielding devices and containers [68]. A UK Government White Paper [70] defines an inventory of materials that may need to be managed through geological disposal: HLW, ILW and some LLW unsuitable for near surface disposal, such as depleted natural and low enriched uranium. In support of the framework, Radioactive Waste Management Ltd has developed a generic Disposal System Safety Case (DSSC) suite of documents for the iterative development of a geological disposal facility. The generic DSSC concludes that the geological disposal facility can be implemented safely in a range of geological environments in the UK.

United States of America

In the USA, exempted users may transfer or dispose of DU-containing products or devices without radiological restriction, and licensed users have to transfer and dispose of products or devices in accordance with regulation 10 CFR 40.51 [70]. Exemption of these devices from nuclear regulation would allow for their disposal as hazardous waste. Users of products or devices containing DU need to (a) qualify for one of the licensing exemptions defined in 10 CFR 40.13 (Unimportant quantities of source material), (b) qualify for one of the general licences in 10 CFR 40.21 through 40.28, or (c) obtain a specific licence from the United States Nuclear Regulatory Commission (USNRC) or an Agreement State. One licensing exemption stated in 10 CFR 40.13 is as follows:

"Natural or depleted uranium metal used as shielding constituting part of any shipping container: *Provided,* That:

(i) The shipping container is conspicuously and legibly impressed with the legend 'CAUTION—RADIOACTIVE SHIELDING—URANIUM'; and
(ii) The uranium metal is encased in mild steel or equally fire resistant metal of minimum wall thickness of one-eighth inch (3.2 mm)" [71].

The Waste Control Specialists (WCS) site in Texas has been approved for DU disposal. An amendment to the WCS licence authorizes the disposal of DU in both the federal and compact waste facilities. The site licensee stated:

"DU is Class A low-level radioactive waste ('LLRW'), which is the most benign of all LLRW, but can increase in radioactivity over time. For that reason, the U.S. Nuclear Regulatory Agency ('NRC') recommends that DU be disposed in containers and buried at least 40 feet below grade. The amendment to WCS' license was unanimously approved because, at WCS, DU will be disposed in the safest and most robust fashion available in the U.S. — encased in concrete at a depth of more than 100 feet."

Technologies are available for processing DU and DU oxides into more environmentally acceptable forms for long term, near surface disposal in a radioactive or hazardous waste disposal site [72].

8. CONCLUSIONS AND STEPS FORWARD

This publication identifies and addresses an emerging area of interest to Member States and intends to raise Member States' awareness of certain issues that have not been recognized until recently. The publication provides general information about DU devices, illustrates a variety of devices where DU is used to shield radiation and radioactive sources, and shows how to identify devices containing DU. It highlights the radiological and chemical challenges posed by DU shielded devices (also called DU devices) and introduces a range of management options, supported by some national examples.

In conclusion, DU presents both chemical and radiological hazards that require suitable controls for the present and the long term. Key points are summarized below.

— The applications of SRSs in medicine, industry, agriculture and research are extremely diverse. Radiation generating and radioactive devices generally incorporate shielding materials, such as DU, to absorb radiation to a level harmless to workers and the public. The identification and characterization of undocumented devices containing DU, as described in Section 4, is essential for keeping a proper inventory and control of this material at a national level.
— Prior to acquiring new devices, the shielding material(s) need to be confirmed to ensure that the user is aware that the management of all acquired DU-containing devices complies with all relevant regulations, including safeguards requirements.
— The DU inventory in a Member State is required to be accurately documented and reported to the regulatory body(ies) in the Member State. The Member State shares the information with the IAEA in the format required by the IAEA Department of Safeguards. The possession, import and export of DU have to be under the control of the Member State.
— Upon request from the Member State, the IAEA can assist and advise Member States on compiling their own comprehensive inventories of DU devices.

— Member States need to have appropriate plans in place for the safe management of DU from disused devices containing DU shielding.
— Each DU device has to be managed (handling, reuse, recycling, transport, storage and disposal) in a controlled manner to ensure the safety of the personnel responsible for working with DU, members of the public and the environment.
— Storage of disused devices containing DU shielding at users' facilities presents safety and security challenges. DU materials (such as empty radiography cameras) in operational storage are at risk of being lost from control. A dedicated national centralized storage facility for DSRSs/waste reduces the likelihood of losing control of DU-containing devices. In any case, specific attention is warranted from the user or the centralized storage operator, as well as the regulators, to provide access control and security, maintain adequate records, determine individual responsibilities and ensure that routine checks (e.g. leak tests) are performed.
— Member States could consider alternatives to DU shielded devices (e.g. tungsten shielded devices), where available.
— It is important that Member States actively pursue strategies for long term DU management, as well as ensuring the sustainability of current practices. In addition, there are a variety of short term actions that can be undertaken, such as:
  - Discussing/negotiating return to suppliers/manufacturers, if they exist; the return of disused devices (containing DU shielding) to the supplier, as envisaged in the Joint Convention (Ref. [13], Article 28), is in principle a good solution. In practice, however, there may be difficulties due to the Member State's legislation, and/or the supplier or manufacturer may have gone out of business.
  - Recycling and/or reusing devices, either domestically or abroad (as discussed in Section 7).
  - Seeking assistance/guidance from the IAEA and other organizations on the near term management of DU.

Appendix I

GENERAL ASPECTS OF RADIOACTIVE WASTE

A conceptual framework of the waste classification scheme, as presented in IAEA General Safety Guide No. GSG-1 [65], is illustrated in Fig. 54. The main parameters used in the classification scheme are the levels of activity content of the waste and the half-lives of the radionuclides contained in the waste, taking into account the hazards posed by different radionuclides and the types of radiation emitted.

The vertical axis represents the activity content of the waste, and the horizontal axis represents the half-lives of the radionuclides contained in the waste. Activity content can range from negligible to very high specific activity. The higher the level of activity content, the greater the need to contain the waste and isolate it from the biosphere. At the lower end of the vertical axis, below clearance levels, the management of the waste can be carried out without consideration of its radiological properties. Activity content covers activity concentration, specific activity and total activity.

For the horizontal axis, the half-lives of the radionuclides contained in the waste can range from short (seconds) to very long (millions of years) time spans. The radiological hazards associated with short lived radionuclides, which generally have a half-life of less than 30 years, are significantly reduced over a few hundred years by radioactive decay. The limitations placed on the activity of waste that can be disposed of in a given facility will depend on the radiological, chemical, physical and biological properties of the waste and on the properties of the particular radionuclides it contains. It should be emphasized that the generic waste classification scheme does not replace the specific safety assessment required for a waste management facility or activity.

A clear distinction needs to be made between a classification scheme and a set of regulatory limits. The development of precise limits has to be carried out within the regulatory framework of licensing or authorizing specific radioactive waste management activities and facilities. It is the responsibility of

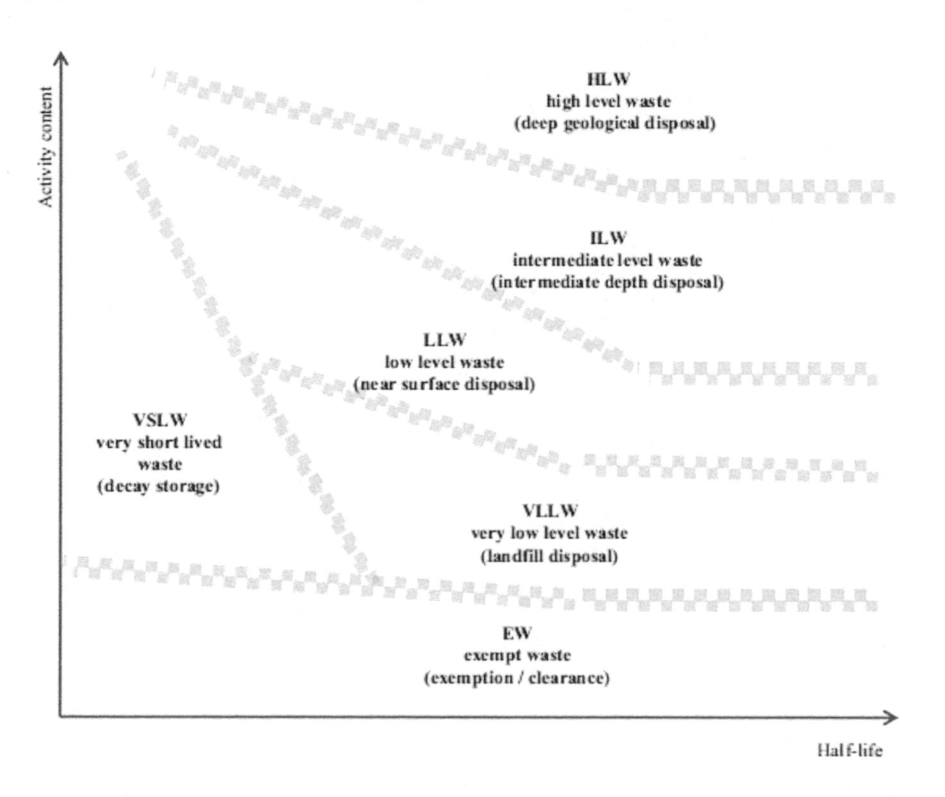

FIG. 54. Waste classification scheme from IAEA General Safety Guide No. GSG-1 [65].

the regulatory body of a Member State to establish actual limits on quantities or concentrations for the classification of radioactive waste. Different Member States use different waste classification schemes. Some Member States have a class of VLLW. In many Member States, further distinction is made on the basis of the half-lives of the radionuclides in the waste, the physical state of the waste and other factors.

With regard to the disposal of DU shields, it should be noted that DU is characterized by very low specific activity, alpha emitting particles and the very long half-life of ^{238}U (millions of years). However, a site-specific safety assessment needs to be carried out to establish the limits, in terms of specific activity, total activity, quantities, and so on, on the DU waste that can be accepted for disposal at the facility.

Appendix II

CASE STUDY: DU SHIELDING CONTROL AND PROTECTION IN FRANCE

II.1 REGULATORY FRAMEWORK

In France, DU is under the control of the High Civil Servant of the Ministry of Industry. The regulatory framework is based on the Code of Defence. This code does not deal with protection of workers and the public against ionizing radiation hazards, nor the safety of nuclear installations, which are governed by the Code for Health and the Environment. The main regulatory documents are:

- Decree no. 2009 1120 dated 17 September 2009, relating to the protection and control of nuclear materials, their facilities and their transportation;
- Order dated 31 May 2011, relating to physical follow up, accounting and physical protection provisions for nuclear material submitted to the declaration regime;
- Order dated 9 June 2011, relating to the implementation of physical follow up and accounting dispositions for nuclear material submitted to authorization;
- Order dated 19 June 2011, relating to physical protection dispositions of facilities holding nuclear materials submitted to authorization.In order to enable France to meet its international obligations, the Institute for Radiation Protection and Nuclear Safety (IRSN) has been commissioned by the Euratom Technical Committee to implement a system for the development of French declarations (information from operators, and development of forms and manuals specifically for declaration purposes).

At the international level, inspections are carried out in France by the IAEA and Euratom. These inspections involve the declaration and monitoring of the movement of nuclear materials (plutonium, uranium and thorium) between countries, as well as declarations regarding the flow and stocks of materials held at a national level that do not relate to materials concerned with national security. International inspections also involve inspections of French facilities by Euratom inspectors and by the IAEA.

At a national level, the protection and inspection of nuclear materials is subject to specific regulations that fall under the Code of Defence and associated regulatory documentation. Given the importance of its nuclear industry and awareness of its responsibilities regarding non-proliferation, France has adopted some of the most comprehensive regulations in the world, covering both civil nuclear materials and those relevant for national security.

The list and amounts of the fusible, fissile and fertile materials covered by French regulations are specified in article R. 1333-1 of Decree No. 2009-1120 of 17 September 2009: plutonium, ^{233}U, uranium enriched in ^{235}U, high enriched, low enriched, natural uranium and DU, thorium, tritium, deuterium and ^{6}Li.

Regarding DU, whatever its physical-chemical form, the limits for quantities held in a facility at a certain time, or during a fiscal year and corresponding regulatory system, are the following:

- Less than 1 kg DU: The operator (holder of devices with DU shielding) is not subject to regulations related to the control and protection of nuclear materials. It may have to comply with other regulations, such as radiation protection regulations. Operators under the declaration regime are called 'declarants'.
- Between 1 kg and 500 kg DU: The operator is submitted to the declaration regime.
- If the operator holds more than 500 kg DU, it shall be granted an authorization (licence) delivered by the Ministry of Industry.

II.2 REQUIREMENTS FOR OPERATORS SUBMITTED TO THE DECLARATION REGIME

Most holders of DU in disused devices containing DU are compelled to the declaration regime. Some need an authorization granted by the Ministry of Industry. The requirements related to the authorization are essentially the same as the requirements for the declaration regime described below, but more stringent, particularly regarding the requirement for a daily accounting report to the IRSN, technical support of the regulatory body.

II.2.1. Annual declaration

A declaration should be established by the operator every year, before 31 January.

Declarations are sent to IRSN, which gathers and centralizes all data within the national boundaries.

The declaration should contain the following data:

— Identification data (company's name, address, etc.) and the name of the DU holder legally responsible;
— Type of activities concerned and location of DU;
— Stock of nuclear materials held on 31 December of the previous year;
— Stock variations that occurred during the previous year, including the identification of senders and receivers;
— Stock and inventory changes of nuclear materials expected for the current year.

II.2.2. Accounting of DU

Declarants should set up a local accounting system based on an inventory book. This book is used to gather chronological records of the various types of inventory changes that occur at the facility, such as reports on materials produced and consumed. The type of book is not stipulated by the order.

II.2.3. Physical inventory

The physical verification of the DU inventory is the most elementary means to reveal a possible loss of control of material or devices. Before fulfilling its annual declaration, the operator must carry out a physical inventory to ensure that the expected amounts of DU and devices, as reflected in the accounting system, are in fact present in the facility.

II.2.4. Physical protection and surveillance

The declaration also has to describe the main features concerning facility layout related to surveillance and physical protection of pieces of DU and devices containing DU.

With respect to the physical protection requirements, these pieces and devices should be kept under lock and key, and keys should be accessible to authorized personnel only.

Alarms and guards are not mandatory, but in some special cases of large DU inventories, an alarm system has been required by the authority.

Enforcement of the declaration regime requirements is ensured by on-site inspections carried out by sworn and accredited inspectors under the authority of the Ministry of Industry.

Appendix III

SPECIFIC REGULATORY REQUIREMENTS FOR STORAGE AND TRANSPORT OF DU IN HUNGARY

III.1. SPECIFIC REQUIREMENTS FROM THE HUNGARIAN SECURITY REGULATIONS FOR PHYSICAL PROTECTION RELATED TO THE STORAGE OF DU

Hungarian security regulations stipulate specific requirements regarding the storage of DU and DU-containing devices, based on the implementation of four security functions: deterrence, detection, delay and response. They are summarized below.

III.1.1 Deterrence function

III.1.1.1. Warning signs

The following information and warning signs shall be applied at the entrance of the storage facility:

— Name of the facility;
— Warning on vehicle entrance rules;
— Warning on unpermitted tools and activities;
— Warning on radiation hazard;
— Warning on entrance hazards and indication of entrance conditions.

III.1.1.2. Artificial barriers (barrier gates, obstructions, chicanes)

Checked entry to the protected facility shall be assured by road signs, if it consists of more than one physical protection zone; otherwise, artificial barriers shall not be applied.

III.1.1.3. Accounting and control requirements

The physical inventory of nuclear and other radioactive materials shall be verified in a documented manner at least once every six months.

III.1.2. Detection function

III.1.2.1. Entrance control system

The entrance control system shall consist of the following elements: lockable doors and limitation of entrance rights.

III.1.3. Delay function

III.1.3.1. Physical protection zones

If the site under protection consists of more than one physical protection zone, then a fence shall be constructed so that:

- — It shall be erected on the border of the building site or around a separated area;
- — It shall be made of steel or plastic wire, with a single foundation, and it shall be a minimum of 2 m high;
- — Fence columns shall be made of steel or concrete;
- — It shall exhibit at least 5 s delay time.

If the site under protection consists of more than one physical protection zone, then the gates located on the fence shall be installed so that:

- — They shall be used as a checkpoint for personal and vehicle access and regress;
- — They shall reach the closed position within the time interval defined in the physical protection plan;
- — They shall be made of steel stronger than the material of the fence;
- — The gates shall provide protection as a vehicle barrier and be at least as high as the fence;
- — They shall be equipped with remotely controlled, motor operated opening equipment, and the grid on the gate shall be made of any material stronger than the fence;
- — Additional gates may provide reserve access routes and support the performance of maintenance actions;
- — The material of these additional gates shall have at least equivalent strength to the material of the fence.

The gates shall be kept closed, and if the physical protection system consists of more than one physical protection zone, they may be opened in the presence of a guard.

The buildings shall comply with the following requirements:

- — The walls can be of lightweight construction, but they shall exhibit limited mechanical resistance;
- — The walls, doors and windows shall exhibit minimum 3 min breakage time against an intruder equipped with common tools.

III.1.3.2. Active movable mechanical barriers and associated locks

The following doors, windows and locks shall be installed so that:

- — The doors shall exhibit at least limited mechanical resistance and a minimum 3 min breakage time against an adversary equipped with common tools;
- — The windows shall have glass of minimum 6 mm width, and the window-protecting grids shall be installed on the inner side of the windows;
- — The locks shall be any traditional locks.

III.1.3.3. Security stores, steel plate cabinets

The security store shall exhibit resistance equivalent to a piece of lockable office furniture.

III.1.4. Response function

III.1.4.1. Internal response forces and on-call police forces

The response shall be performed by internal or external response forces.

III.2. SPECIFIC REQUIREMENTS FROM THE HUNGARIAN SECURITY REGULATIONS FOR THE TRANSPORT OF DU

As with storage, Hungarian regulatory requirements for the transport of DU and DU-containing devices are based on the implementation of four security functions: deterrence, detection, delay and response. They are summarized below.

III.2.1. Deterrence function

III.2.1.1. Warning placards and signs

Signs indicating radiation hazard and warnings regarding prohibited tools and activities shall be applied (i) in the case of a closed transport vehicle, on the door of the cargo compartment; and (ii) in the case of an open transport vehicle, on the outside of the vehicle, with the exception of the surface of the transported consignment.

III.2.1.2. Communication

Continuous, reliable mobile communication shall be provided with the licensee during the transport.

III.2.1.3. Knowledge of physical protection

Training in physical protection is mandatory every three years for those who take part in transport activities.

III.2.1.4. Open vehicle

Consignments weighing more than 500 kg may be transported on an open vehicle if the physical protection equipment is able to provide reliable protection against unauthorized removal.

III.2.2. Detection function

With respect to the detection security function, the requirements are that a signalling system should be used to detect unauthorized access and attack, as follows. The signalling system shall be activated if any of the following occur: opening, breaking, cutting or disassembling the sensors.

III.2.3. Delay function

The delay function concerns the transport vehicle and sheet metal cabinet doors, as follows:

- The doors of the transport vehicle or the doors of the sheet metal cabinet shall exhibit limited mechanical resistance against hand tools;
- The transport vehicle or sheet metal cabinet doors shall exhibit resistance against an intruder equipped with traditional hand tools for a minimum of 3 min.

The delay function also relates to the chassis of the transport vehicle and the storage plate cabinet, as follows:

- The chassis of the transport vehicle or sheet metal cabinet shall exhibit resistance equivalent to that of a piece of lockable office furniture.

III.2.4. Response function

The primary and alternative routes, as well as the stops during transport, shall be planned so that the response can be executed by escort personnel and any additional on-call external response forces.

REFERENCES

[1] INTERNATIONAL ATOMIC ENERGY AGENCY, IAEA Safety Glossary, Terminology Used in Nuclear Safety and Radiation Protection, 2018 Edition, IAEA, Vienna (2019).

[2] INTERNATIONAL ATOMIC ENERGY AGENCY, IAEA Safeguards Glossary, International Nuclear Verification Series No. 3, 2001 Edition, IAEA, Vienna (2002).

[3] INTERNATIONAL ATOMIC ENERGY AGENCY, Identification of Radioactive Sources and Devices, IAEA Nuclear Security Series No. 5, IAEA, Vienna (2007).

[4] INTERNATIONAL ATOMIC ENERGY AGENCY, Review of Sealed Source Designs and Manufacturing Techniques Affecting Disused Source Management, IAEA TECDOC 1690, IAEA, Vienna (2012).

[5] INTERNATIONAL ATOMIC ENERGY AGENCY, Management of Spent High Activity Radioactive Sources (SHARS), IAEA TECDOC 1301, IAEA, Vienna (2002).

[6] INTERNATIONAL ATOMIC ENERGY AGENCY, Management of Disused Sealed Radioactive Sources, IAEA Nuclear Energy Series No. NW T 1.3, IAEA, Vienna (2014).

[7] INTERNATIONAL ATOMIC ENERGY AGENCY, Management of Disused Long Lived Sealed Radioactive Sources (LLSRS), IAEA TECDOC 1357, IAEA, Vienna (2003).

[8] INTERNATIONAL ATOMIC ENERGY, Categorization of Radioactive Sources, IAEA Safety Standards Series No. RS G 1.9, IAEA, Vienna (2005).

[9] INTERNATIONAL ATOMIC ENERGY AGENCY, Guidance on the Management of Disused Radioactive Sources, IAEA/CODEOC/MGT DRS/2018, IAEA, Vienna (2018).

[10] INTERNATIONAL ATOMIC ENERGY AGENCY, Predisposal Management of Radioactive Waste, IAEA Safety Standards Series No. GSR Part 5, IAEA, Vienna (2009).

[11] INTERNATIONAL ATOMIC ENERGY AGENCY, Disposal of Radioactive Waste, IAEA Specific Safety Requirements No. SSR 5, IAEA, Vienna (2011).

[12] INTERNATIONAL ATOMIC ENERGY AGENCY, Code of Conduct on the Safety and Security of Radioactive Sources, IAEA/CODEOC/2004, IAEA, Vienna (2004).

[13] INTERNATIONAL ATOMIC ENERGY AGENCY, Joint Convention on the Safety of Spent Fuel Management and on the Safety of Radioactive Waste Management, INFCIRC/546, IAEA, Vienna (1997).

[14] INTERNATIONAL ATOMIC ENERGY AGENCY, Disposal Options for Disused Radioactive Sources, Technical Reports Series No. 436, IAEA, Vienna (2005).

[15] INTERNATIONAL ATOMIC ENERGY AGENCY, Borehole Disposal Facilities for Radioactive Waste, IAEA Safety Standards Series No. SSG 1, IAEA, Vienna (2009).

[16] INTERNATIONAL ATOMIC ENERGY AGENCY, Safeguards Implementation Guide for States with Small Quantities Protocols, IAEA Services Series No. 22, IAEA, Vienna (2013).

[17] HERTZLER, T.J., NISHIMOTO, D.D., OTIS, M.D., Depleted Uranium Disposal Options Evaluation, Tech. Rep. EGG MS 11297, Science Applications International Corporation, Idaho Falls, ID (1994).

[18] US ENVIRONMENTAL PROTECTION AGENCY, Depleted Uranium — Technical Brief, EPA 402 R 06 011, USEPA, Washington, DC (2006), https://www.epa.gov/sites/production/files/2015 05/documents/402 r 06 011.pdf

[19] US NUCLEAR REGULATORY COMMISSION, Background information on depleted uranium, USNRC, Washington, DC (2017),
http://www.nrc.gov/about nrc/regulatory/rulemaking/potential rulemaking/uw streams/bg info du.html

[20] US NUCLEAR REGULATORY COMMISSION, Commission Paper SECY 08 0147, Analysis of Depleted Uranium Disposal, USNRC, Washington, DC (2008).

[21] CHEM NUCLEAR SYSTEMS, Interim Site Stabilization and Closure Plan for the Barnwell Low level Radioactive Waste Disposal Facility — 2005 Closure Plan, Chem Nuclear Systems, Barnwell, SC (2005).

[22] WORLD NUCLEAR ASSOCIATION, Uranium and Depleted Uranium, World Nuclear Association, London (2016).

[23] EUROPEAN COMMISSION, FOOD AND AGRICULTURE ORGANIZATION OF THE UNITED NATIONS, INTERNATIONAL ATOMIC ENERGY AGENCY, INTERNATIONAL LABOUR ORGANIZATION, OECD NUCLEAR ENERGY AGENCY, PAN AMERICAN HEALTH ORGANIZATION, UNITED NATIONS ENVIRONMENT PROGRAMME, WORLD HEALTH ORGANIZATION, Radiation Protection and Safety of Radiation Sources: International Basic Safety Standards, IAEA Safety Standards Series No. GSR Part 3, IAEA, Vienna (2014).

[24] INTERNATIONAL ATOMIC ENERGY AGENCY, Radiation Safety in Industrial Radiography, IAEA Safety Standards Series No. SSG 11, IAEA, Vienna (2011).

[25] HEALTH PHYSICS SOCIETY, Answer to Question #8929 Submitted to 'Ask the Experts', HPS, Herndon, VA (2014), hps.org/publicinformation/ate/q8929.html

[26] GONZALEZ, A., Security of radioactive sources: the evolving new international dimensions, IAEA Bull. 43 4 (2001) 39.

[27] AMERICAN NATIONAL STANDARDS INSTITUTE, Safe Design and Use of Self Contained, Dry Source Storage Gamma Irradiators (Category I), Standard N433.1, National Bureau of Standards Handbook, ANSI, Washington, DC (1978).

[28] INTERNATIONAL ATOMIC ENERGY AGENCY, Radiation Protection and Safety in Industrial Radiography, IAEA Safety Reports Series No. 13, IAEA, Vienna (1999).

[29] INTERNATIONAL ATOMIC ENERGY AGENCY, Regulations for the Safe Transport of Radioactive Material (2018 Edition), IAEA Safety Standards Series No. SSR 6 (Rev. 1), IAEA, Vienna (2018).

[30] MATVEEV, V.Z., et al., "Design of transport casks with depleted uranium gamma shield and advanced safety" (Proc. Int. High Level Radioactive Waste Management Conf. (IHLRWM), Las Vegas, 30 April – 4 May 2006), ANS, La Grange Park, IL (2006),
http://citeseerx.ist.psu.edu/viewdoc/download?doi=10.1.1.513.7839&rep=rep1&type=pdf

[31] FERRADA, J.J., DOLE, L.R., ERMICHEV, S.G., "Analyses of US and Russian Federation depleted uranium concrete/steel transport containers" (Proc. Int. High Level Radioactive Waste Management Conf., Las Vegas, 30 April – 4 May 2006), ANS, La Grange Park, IL (2006),
https://pdfs.semanticscholar.org/3cf4/8efe02dbc35bf367795644fe4769f061cfe7.pdf

[32] EUROPEAN ATOMIC ENERGY COMMUNITY, Commission Regulation (Euratom) No. 302/2005 of 8 February 2005 on the application of Euratom safeguards (2013.07.01), OJ L54 (2013) 1,
http://data.europa.eu/eli/reg/2005/302/2013 07 01

[33] INTERNATIONAL ATOMIC ENERGY AGENCY, Disused Sealed Radioactive Sources Network — DSRSNet, IAEA, Vienna (2022),
https://nucleus.iaea.org/sites/connect/DSRSpublic/Pages/default.aspx

[34] INTERNATIONAL ATOMIC ENERGY AGENCY, Directory of National Competent Authorities' Approval Certificates for Package Design, Special Form Material and Shipment of Radioactive Material, 2004 Edition, IAEA TECDOC 1424, IAEA, Vienna (2004).

[35] INTERNATIONAL ATOMIC ENERGY AGENCY, The Structure and Content of Agreement Between the Agency and States Required in Connection with the Treaty on the Non Proliferation of Nuclear Weapons, INFCIRC/153 (Corrected), IAEA, Vienna (1972).

[36] INTERNATIONAL ATOMIC ENERGY AGENCY, The Standard Text of Safeguards Agreements in Connection with the Treaty on the Non Proliferation of Nuclear Weapons, GOV/INF/276, IAEA, Vienna (1974).

[37] INTERNATIONAL ATOMIC ENERGY AGENCY, Guidance for States Implementing Comprehensive Safeguards Agreements and Additional Protocols, IAEA Services Series No. 21, IAEA, Vienna (2016).

[38] INTERNATIONAL ATOMIC ENERGY AGENCY, Nuclear Material Accounting Handbook, IAEA Services Series No. 15, IAEA, Vienna (2008).

[39] INTERNATIONAL ATOMIC ENERGY AGENCY, Governmental, Legal and Regulatory Framework for Safety, IAEA Safety Standards Series No. GSR Part 1 (Rev. 1), IAEA, Vienna (2016).

[40] INTERNATIONAL ATOMIC ENERGY AGENCY, Security of Radioactive Material in Use and Storage and of Associated Facilities, IAEA Nuclear Security Series No. 11 G (Rev.1), IAEA, Vienna (2019).

[41] US DEPARTMENT OF ENERGY, DOE STD 1136 2017, Good Practices for Occupational Radiological Protection in Uranium Facilities, USDOE, Washington, DC (2017).

[42] EUROPEAN ATOMIC ENERGY COMMUNITY, Council Directive 2013/59/EURATOM of 5 December 2013 on laying down basic safety standards for protection against the dangers arising from exposure to ionising radiation, and repealing Directives 89/618/Euratom, 90/641/Euratom, 96/29/Euratom, 97/43/Euratom and 2003/122/Euratom, OJ L13 (2014) 1.

[43] INTERNATIONAL ATOMIC ENERGY AGENCY, Optimization of Radiation Protection in the Control of Occupational Exposure, Safety Reports Series No. 21, IAEA, Vienna (2002).

[44] INTERNATIONAL ATOMIC ENERGY AGENCY, Nuclear Security Recommendations on Physical Protection of Nuclear Material and Nuclear Facilities (INFCIRC/225/Revision 5), Nuclear Security Recommendations, IAEA Nuclear Security Series No. 13, IAEA, Vienna (2011).

[45] INTERNATIONAL ATOMIC ENERGY AGENCY, Nuclear Security Recommendations on Radioactive Material

and Associated Facilities, Nuclear Security Recommendations, IAEA Nuclear Security Series No. 14, IAEA, Vienna (2011).
[46] INTERNATIONAL ATOMIC ENERGY AGENCY, Physical Protection of Nuclear Material and Nuclear Facilities (Implementation of INFCIRC/225/Revision 5), Implementing Guides, IAEA Nuclear Security Series No. 27 G, IAEA, Vienna (2018).
[47] INTERNATIONAL ATOMIC ENERGY AGENCY, Advisory Material for the IAEA Regulations for the Safe Transport of Radioactive Material (2012 Edition), IAEA Safety Standards Series No. SSG 26, IAEA, Vienna (2012).
[48] INTERNATIONAL ATOMIC ENERGY AGENCY, Schedules of Provisions of the IAEA Regulations for the Safe Transport of Radioactive Material (2018 Edition), IAEA Safety Standards Series No. SSG 33 (Rev. 1), IAEA, Vienna (2021).
[49] INTERNATIONAL ATOMIC ENERGY AGENCY, Preparedness and Response for a Nuclear or Radiological Emergency Involving the Transport of Radioactive Material, IAEA Safety Standards Series No. SSG 65, IAEA, Vienna (2022).
[50] INTERNATIONAL ATOMIC ENERGY AGENCY, Radiation Protection Programmes for the Transport of Radioactive Material, IAEA Safety Standards Series No. TS G 1.3, IAEA, Vienna (2007).
[51] INTERNATIONAL ATOMIC ENERGY AGENCY, T he Management System for the Safe Transport of Radioactive Material, IAEA Safety Standards Series No. TS G 1.4, IAEA, Vienna (2008).
[52] INTERNATIONAL ATOMIC ENERGY AGENCY, Compliance Assurance for the Safe Transport of Radioactive Material, IAEA Safety Standards Series No. TS G 1.5, IAEA, Vienna (2009).
[53] INTERNATIONAL ATOMIC ENERGY AGENCY, Security of Radioactive Material in Transport, Implementing Guides, IAEA Nuclear Security Series No. 9 G (Rev. 1), IAEA, Vienna (2020).
[54] INTERNATIONAL ATOMIC ENERGY AGENCY, Security of Nuclear Material in Transport, Implementing Guides, IAEA Nuclear Security Series No. 26 G, IAEA, Vienna (2015).
[55] INTERNATIONAL ATOMIC ENERGY AGENCY, Leadership, Management and Culture for Safety in Radioactive Waste Management, IAEA Safety Standards Series No. GSG 16, IAEA, Vienna (2022).
[56] INTERNATIONAL ATOMIC ENERGY AGENCY, Predisposal Management of Radioactive Waste from the Use of Radioactive Material in Medicine, Industry, Agriculture, Research and Education, IAEA Safety Standards Series No. SSG 45, IAEA, Vienna (2019).
[57] CANADIAN NUCLEAR SAFETY COMMISSION, Fact sheet: Depleted uranium: the Canadian regulator's perspective, CNSC, Ottawa, ON (2009), http://nuclearsafety.gc.ca/eng/resources/fact sheets/depleted uranium perspective.cfm
[58] INTERNATIONAL ATOMIC ENERGY AGENCY, Predisposal Management of Radioactive Waste from Nuclear Power Plants and Research Reactors, IAEA Safety Standards Series No. SSG 40, IAEA, Vienna (2016).
[59] INTERNATIONAL ATOMIC ENERGY AGENCY, Storage of Radioactive Waste, IAEA Safety Standards Series No. WS G 6.1, IAEA, Vienna (2006).
[60] INTERNATIONAL ATOMIC ENERGY AGENCY, Classification of Radioactive Waste, General Safety Guide No. GSG 1, IAEA, Vienna (2009).
[61] INTERNATIONAL ATOMIC ENERGY AGENCY, Near Surface Disposal Facilities for Radioactive Waste, IAEA Safety Standards Series No. SSG 29, IAEA, Vienna (2014).
[62] INTERNATIONAL ATOMIC ENERGY AGENCY, Geological Disposal Facilities for Radioactive Waste, IAEA Safety Standards Series No. SSG 14, IAEA, Vienna (2011).
[63] INTERNATIONAL ATOMIC ENERGY AGENCY, Design Principles and Approaches for Radioactive Waste Repositories, IAEA Nuclear Energy Series No. NW T 1.27, IAEA, Vienna (2020).
[64] INTERNATIONAL ATOMIC ENERGY AGENCY, Underground Disposal Concepts for Small Inventories of Intermediate and High Level Radioactive Waste, IAEA TECDOC 1934, IAEA, Vienna (2020).
[65] INTERNATIONAL ATOMIC ENERGY AGENCY, Policies and Strategies for Radioactive Waste Management, IAEA Nuclear Energy Series No. NW G 1.1, IAEA, Vienna (2009).
[66] ORGANISATION FOR ECONOMIC CO OPERATION AND DEVELOPMENT NUCLEAR ENERGY AGENCY, Low Level Radioactive Waste Repositories: An Analysis of Costs, OECD, Paris (1999).
[67] FRENCH NUCLEAR SAFETY AUTHORITY (ASN), French National Plan for the Management of Radioactive Materials and Waste for 2016–2018, ASN, Montrouge, France (2017), http://www.french nuclear safety.fr/Information/Publications/Others ASN reports/French National Plan for the Management of Radioactive Materials and Waste for 2016 2018
[68] GALSON SCIENCES, Review of UK and Overseas Depleted, Natural and Low Enriched Uranium Management, Integrated Project Team on Uranium: Phase 2, 1207 INT 6 1, Version 1.3, Galson, Oakham, UK (2014).

https://www.nda.gov.uk/publication/review of uk and overseas depleted natural and low enriched uranium management/
[69] DEPARTMENT OF ENERGY AND CLIMATE CHANGE, Implementing Geological Disposal: A Framework for the Long Term Management of Higher Activity Radioactive Waste, URN 14D/235, Department of Energy and Climate Change, London, 2014.
[70] US NUCLEAR REGULATORY COMMISSION, Code of Federal Regulations, Title 10, Part 40, Domestic Licensing of Source Material, USNRC, Washington, DC (2021),
http://www.nrc.gov/reading rm/doc collections/cfr/part040/
[71] WASTE CONTROL SPECIALISTS, Amendment will give WCS ability to dispose of depleted uranium, WCS, Dallas, TX (2014),
http://www.wcstexas.com/wp content/uploads/2015/04/Aug_22_2014 2.pdf
[72] US NUCLEAR REGULATORY COMMISSION, Frequently asked questions about depleted uranium deconversion facilities, USNRC, Washington, DC (2015),
https://www.nrc.gov/materials/fuel cycle fac/ur deconversion/faq depleted ur decon.html

CONTENTS OF THE ANNEXES

NATIONAL REPORTS: EXAMPLES OF NATIONAL EXPERIENCES

There are differences in regulations regarding the possession, control, inventory and disposition (reuse, recycle, import, export, storage or disposal) of DU among Member States. There are also differences in the nuclear security aspects of possession of DU among Member States.

The national reports in the following annexes were prepared by Member State representatives in an IAEA Technical Meeting (19–23 August 2019, Vienna, Austria), according to a brief questionnaire provided by one of the meeting's consultants. The national reports provide snapshots of the status of the management of DU devices for the Member States shown.

The questionnaire asked participants for the following information:

(1) What types of devices, equipment and containers with DU as shielding material do you have in your country?
(2) Are you collecting them in a centralized storage facility (CSF), disposal facility or keeping them at users' premises?
(3) Do you have any safety regulations for the management of this material? Please describe relevant experiences in your country on DU devices storage, transport, disposal, return to manufacturer, etc.);
(4) Do you have any security regulations? Experiences?
(5) Do you have any experience in recycling (in or outside your country) devices containing DU?
(6) Have you identified any problems or need for assistance from the IAEA?

The national reports do not necessarily describe best practices. Rather, they reflect a wide variety of national legislation and policies, social and economic conditions, and nuclear programmes. Although the information presented is not intended to be exhaustive, readers are encouraged to evaluate the applicability of the information in these annexes to specific projects of their concern. These national reports reflect the experience and views of their contributors and, although generally consistent with the main text, are not intended as specific guidance.

Annex I

MEMBER STATE EXPERIENCE: AUSTRALIA

I–1. TYPES OF DEVICES THAT CONTAIN DU

The Australian DU inventory comprises primarily old radioactive source transport containers, shielding and collimators from teletherapy heads, industrial gauges, source changers for industrial radiography and gamma irradiators (Fig. I–1). The Australian Nuclear Science and Technology Organisation (ANSTO) was the only manufacturer of SRSs in Australia, ceasing manufacturing operations in 2005.

I–2. STORAGE AND DISPOSAL FACILITIES FOR DU

DU is considered a nuclear material and is stored at the premises of the licensed/permitted user until a final disposition pathway is determined. The preferred storage (interim) for DU is in a CSF, for countries with a DU or radioactive waste inventory that justifies the cost and operation of such a facility.

In Australia, DU is stored in a number of locations. These include a dedicated engineered nuclear material storage facility at the ANSTO site located on the outskirts of Sydney (Fig. I–2), the Australian Defence Department, the Esk radioactive waste storage facility located outside of Brisbane (Fig. I–3) and several users' premises, including universities and hospitals across Australia.

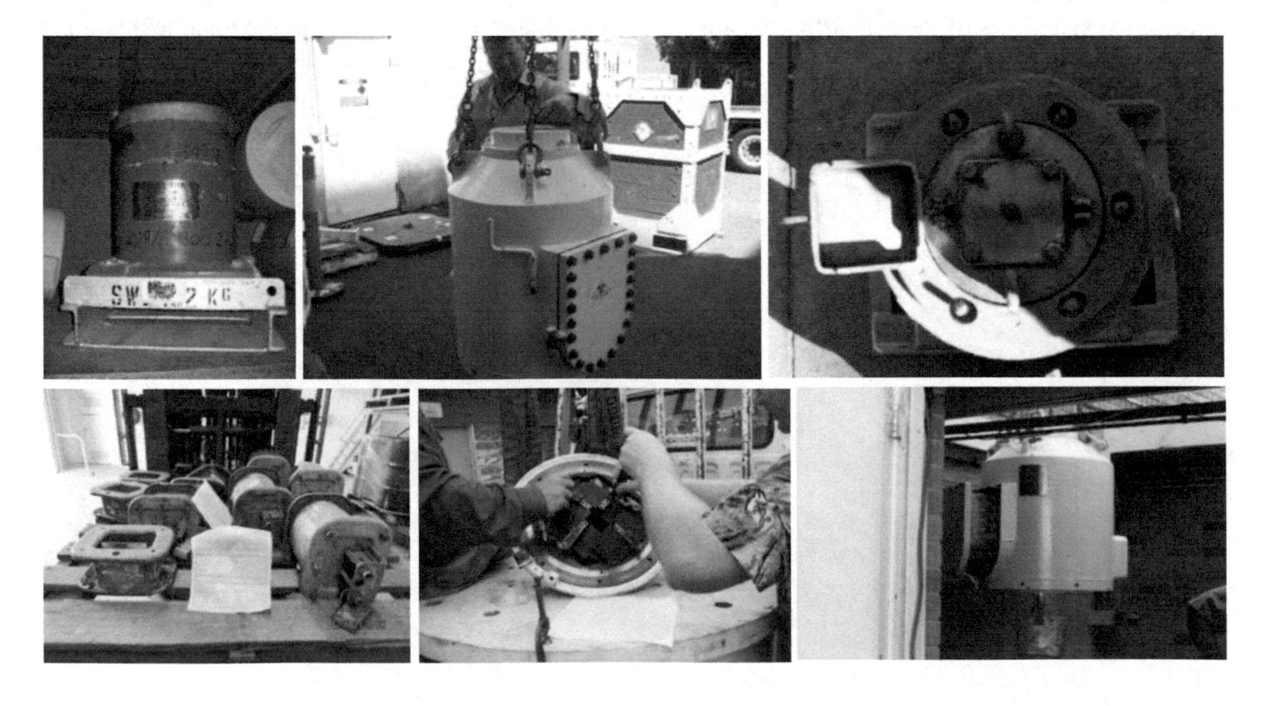

FIG. I–1. Radioactive source transport containers and transport devices, industrial gauges, collimators and gamma irradiators containing DU.

FIG. I–2. Nuclear materials store at ANSTO (NSW).

FIG. I–3. Esk radioactive waste storage facility (Queensland).

I–3. REGULATIONS FOR SAFETY AND SECURITY

Devices containing DU are considered as radioactive sources or material with oversight by several regulators. These include State radiation protection agencies (for all non-Commonwealth entities, e.g. hospitals and mines) and the Australian Radiation Protection and Nuclear Safety Agency (ARPANSA), which is the Commonwealth Government's primary authority on radiation protection and nuclear safety under the Australian Radiation Protection and Nuclear Safety Act 1998.

The Australian transport safety and security regulatory framework is governed by Commonwealth, State and Territory legislations. There are 11 competent authorities in Australia, including three Commonwealth authorities, six State and two Territory authorities. The IAEA Regulations for the Safe Transport of Radioactive Material are applied through the ARPANSA Code of Practice for the Safe Transport of Radioactive Material (2008) by road, rail and waterways not covered by marine legislation. All states and territories apply this transport code through their regulatory systems.

The security of radioactive material (nuclear material) is governed by two Commonwealth agencies, namely ARPANSA and the Australian Safeguards and Non-proliferation Office (ASNO). ASNO regulates the security of nuclear material, including uranium, thorium and plutonium, through the Nuclear Non-Proliferation (Safeguards) Act.

When transporting DU, shipments must comply with the provisions of the national Dangerous Goods Transport Act (the European Agreement Concerning the International Carriage of Dangerous Goods by Road (ADR) is implemented in the national legislation). For portable devices, the devices or containers must have a valid transport certificate.

ASNO issues permits for the possession and transport of all nuclear material, including DU. It also keeps an updated database of all nuclear material present in Australia, conducts regular inspections of all permit holders, facilitates IAEA inspections of Australian facilities, conducts outreach, and provides training and education on nuclear regulation and physical protection.

During the useful lifetime of the device or the container that contains DU, the user must keep it in temporary storage at their premises. When the device or the container that contains DU is disused, the user has several options:

— Return it to the manufacturer/supplier;
— Transfer it to another authorized user;
— Send it for recycling.

I–4. EXPERIENCE REGARDING RETURN TO MANUFACTURER/SUPPLIER

Return of the DU to the original supplier is possible but can be expensive and difficult (e.g. because of the age of the transport containers or the absence of associated drawings/technical data). Recycling options are also available in the USA and Sweden.

Since 1958, ANSTO has amassed considerable amounts of legacy nuclear material, which was surplus to operational requirements. To promote efficiency, reduce regulatory compliance burden and save on future disposition costs, between 2008 and 2011 ANSTO shipped surplus stocks of nuclear material to the USA for recycling and eventual reuse in peaceful applications.

Surplus nuclear material shipped included Pu-Be sources, nuclear grade graphite, tritiated heavy water and a single shipment of DU. The DU was shipped to Manufacturing Sciences Corporation, located in Oak Ridge, Tennessee, for recycling purposes.

The DU shipment (Fig. I–4) included 10.29 tonnes of mixed DU, a teletherapy head (2984 kg) containing DU shielding and various consolidated materials (containing mixed uranium samples). The DU was surplus to needs from discontinued projects. The majority of the DU was in the form of metal (solid or powder). Activities ranged from 0.71 GBq to 28.89 GBq with a transport index of 0.4.

The DU was originally planned to be exported by sea freight but was finally transported by air using a 747 cargo plane. The original sea freight packaging (single ISO shipping container) had to be unloaded and the DU repackaged to conform to air shipment regulations.

I–5. HAVE YOU IDENTIFIED ANY PROBLEMS OR NEED FOR ASSISTANCE FROM THE IAEA?

No. Australia is happy to share its experience with other Member States in relation to managing DU material.

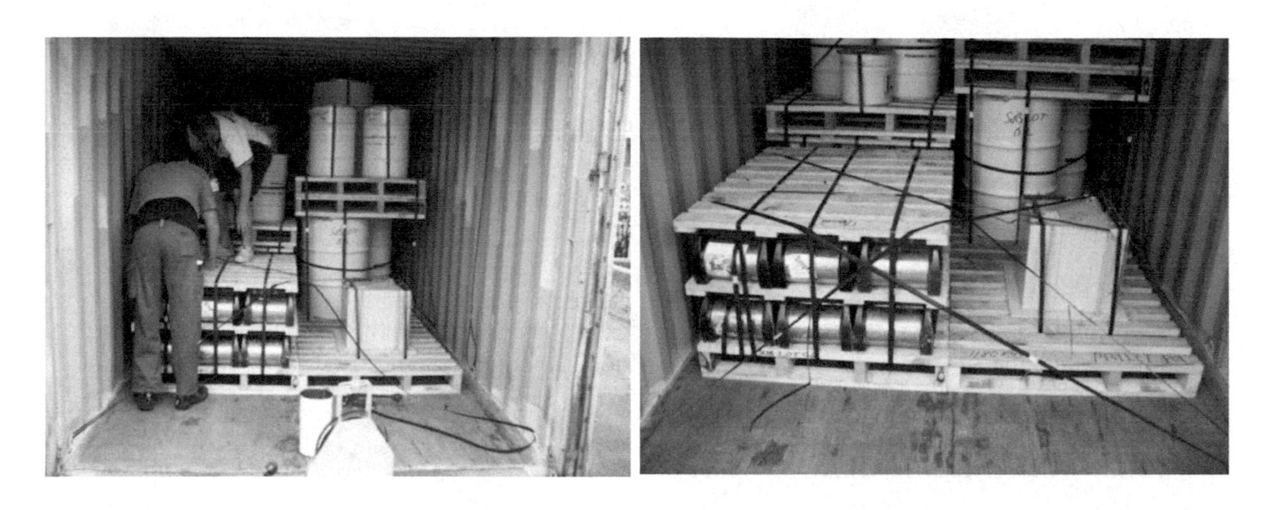

FIG. I–4. DU shipment being packed for shipment to the USA for recycling.

Annex II

MEMBER STATE EXPERIENCE: AZERBAIJAN

II–1. TYPES OF DEVICES THAT CONTAIN DU

Currently, the main types of radioisotope equipment in Azerbaijan that contain DU as biological shielding include industrial gamma radiography and medical radiotherapy devices (Fig. II–1).

II–2. STORAGE AND DISPOSAL FACILITIES FOR DU

A centralized facility for the temporary and long term storage of radioactive waste and DSRSs is in operation. This allows for centralized control of DU materials and devices. The storage facility is located 33 km from the city of Baku, occupying an area of 6 ha (Fig. II–2).

Collection, transportation, treatment, conditioning and storage of radioactive waste and disused radioactive sources in Azerbaijan is managed by the Specialized Enterprises 'Isotope' unit, presently residing under the authority of the Ministry of Emergency Situations.

II–3. REGULATIONS FOR SAFETY AND SECURITY

Azerbaijan has a law on radiation safety, as well as regulatory documents that require authorization of activities related to the use, storage, transportation, processing and other uses of radioactive substances.

Import, export and transit of nuclear and other radioactive materials is authorized through approvals issued by the Cabinet of Ministers following appropriate checks and approval by the relevant authorities. There are national legislative acts in place that define the rules and regulations for the authorization and control of export, import and transit of materials subject to such control. A new law on radiation protection and nuclear security has been drafted and places special emphasis on the requirements for the control and accounting of nuclear materials and the implementation of appropriate safeguards measures.

On 18 March 2016, Azerbaijan signed the 2005 Amendment to the Convention on Physical Protection of Nuclear Materials by the decision of the Milli Mejlis (National Parliament of Azerbaijan)

FIG. II–1. Examples of devices containing DU shielding in Azerbaijan.

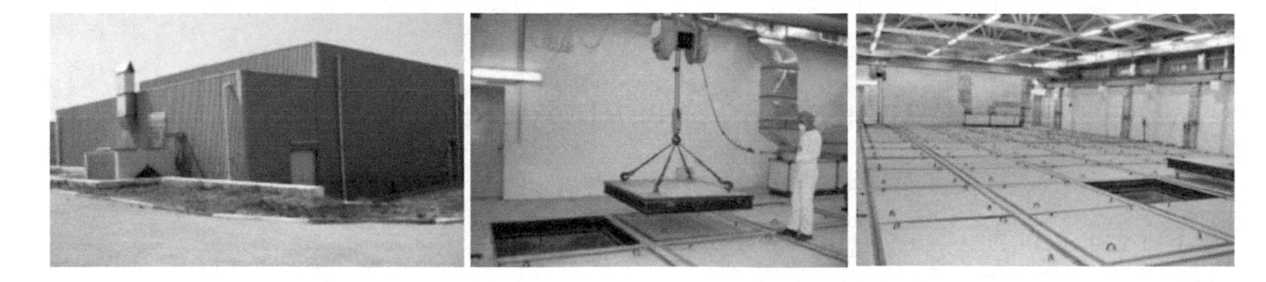

FIG. II–2. Central storage facility for disused sources of low and medium activity radioactive waste.

and Law No. 178-VQ. This was signed by the President of Azerbaijan, with the special declaration related to its occupied territories.

Azerbaijan pays significant attention to the strengthening of the legal framework and regulatory infrastructure in all areas of safety and security. This attention is directed, in particular, to the definition of the requirements of nuclear security and the physical protection of potentially dangerous and specialized facilities, as well as nuclear and other radioactive materials.

A new law on radiation protection and nuclear security has been drafted that addresses the legal framework for the physical protection of nuclear and radioactive materials. The new law is in accordance with obligations under the international treaties and agreements to which Azerbaijan is a party.

II–4. EXPERIENCE REGARDING RETURN TO MANUFACTURER/SUPPLIER

The authorization system for temporary import of radioactive sources and nuclear materials requires their re-export upon completion of use.

Azerbaijan has some experience in exporting radioisotope equipment containing DU, as well as in the manufacturing, decommissioning, disassembling, transporting and preparing for storage of obsolete equipment containing DU.

II–5. EXPERIENCE REGARDING REUSE AND RECYCLING

No experience in these fields.

II–6. HAVE YOU IDENTIFIED ANY PROBLEMS OR NEED FOR ASSISTANCE FROM THE IAEA?

The design of radioactive source equipment is not all the same. This means that special equipment and an individual approach is required for each case of dismantling and decommissioning. In particular, the recycling process requires the presence of appropriate technologies and enterprises, as well as the corresponding infrastructure, which is not available in most of the IAEA Member States. As such, there is a need to implement an internationally unified approach to the issue, or to transfer the recycling/reprocessing of DU-containing shielding materials to authorized organizations.

Annex III

MEMBER STATE EXPERIENCE: BULGARIA

III–1. TYPES OF DEVICES THAT CONTAIN DU

In Bulgaria, there are different types of devices, equipment and containers that contain DU as a shielding material, for example:

- Industrial radiography devices (^{75}Se, ^{192}Ir), crawlers, transport and source changer containers (see Fig. III–1 and summary in Table III–21);
- Teletherapy machines with ^{60}Co with ~6732 kg DU (see Fig. III–2).

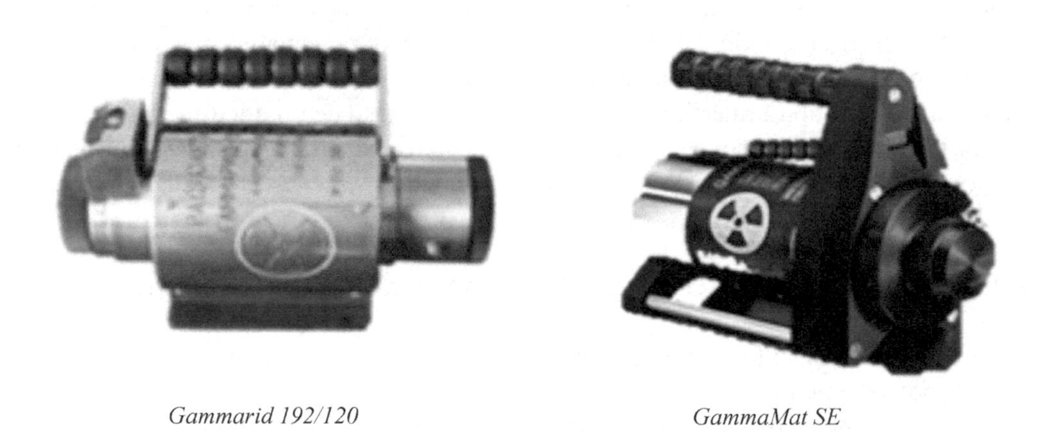

Gammarid 192/120 *GammaMat SE*

FIG. III–1. Two types of industrial radiography devices.

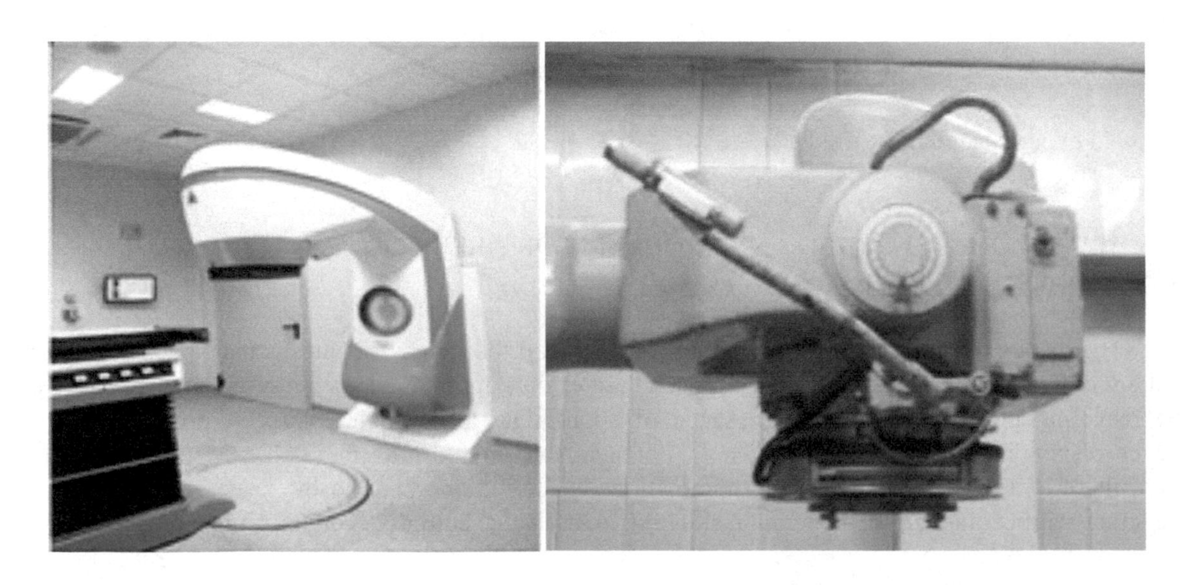

FIG. III–2. Teletherapy machine and head (Rokus-M) containing DU.

III–2. STORAGE AND DISPOSAL FACILITIES FOR DU

At present, there is no CSF for disused devices containing DU shielding in Bulgaria. All disused sources with DU shielding are temporarily stored in separate engineered facilities under conditions specified in the licences and permits issued by the nuclear regulatory agency.

All licensees and permit holders are obliged to deliver their disused sources, including DU shielding, to State Enterprise Radioactive Waste (SERAW) after they have been declared as radioactive waste.

SERAW currently has about 15 241 kg of DU in temporary storage (see Fig. III–3).

III–3. REGULATIONS FOR SAFETY AND SECURITY

The Bulgarian legislative framework has adopted and follows the IAEA and ICRP recommendations and is based on the European legislation in this area. The main normative acts relating to safety management of DU shielding materials include the following:

— Act on safe use of nuclear energy;
— Regulation on radiation protection;
— Regulation on radiation protection during work activities with radiation defectoscopes;
— Regulation on the conditions and procedure for transport of radioactive material;
— Regulation on the safe management of radioactive waste;
— Regulation on the procedure for issuing licences and permits for the safe use of nuclear energy.

TABLE III–1. DIFFERENT TYPES OF INDUSTRIAL RADIOGRAPHY DEVICES

Device name	Device weight (kg)
Gammarid 25	16
Gammarid 25M	16
Gammarid 192/120	30 (~16 kg)
Gammarid 192/120 MD	27 (~16 kg)
Gammavolt SU 50	15
GammaMat TI	12
GammaMat R30	12
GammaMat SE	7 (~2.7 kg)
Exertus dual 60/120	14
GammaMat M6	27
GammaMat M18	27

FIG. III–3. Hot cells (Multitest) and storage facilities for radiography devices, and concrete containers for storage of teletherapy heads (held at SERAW).

The main normative acts concerning the nuclear security of DU shielding materials include:

- Act on safe use of nuclear energy;
- Regulation on the application of the safeguards under the NPT;
- Regulation for the provision of physical protection of nuclear facilities, nuclear material and radioactive material.

III–4. EXPERIENCE REGARDING RETURN TO MANUFACTURER/SUPPLIER

It is important that countries that have previous experience in the recycling of devices containing DU can share these experiences with other Member States in relation to the safe and economic management of DU material.

III–5. HAVE YOU IDENTIFIED ANY PROBLEMS OR NEED FOR ASSISTANCE FROM THE IAEA?

Specific requirements and detailed recommendations should be developed concerning safety management of DU shielding materials, including appropriate options for:

- Short term storage or long term storage and disposal;
- Transport, transit and trans-shipment;
- Reusing or recycling;
- Return to supplier.

A graded approach needs to be implemented, and international and regional cooperation might be continued in this area.

Annex IV

MEMBER STATE EXPERIENCE: CANADA

IV–1. TYPES OF DEVICES THAT CONTAIN DU

These include radiographic exposure devices and transport containers such as industrial gauges, GammaMat radiography cameras, source changers, Tech Ops transport containers, calibration devices and collimators (Fig. IV–1).

Calibration devices are used for calibrating radiation detection equipment. Collimators are used in ^{60}Co devices.

IV–2. STORAGE AND DISPOSAL FACILITIES FOR DU

There is no current CSF in Canada.

DU container shielding material is stored at the user's premises in licensed facilities with appropriately secured storage vaults (Fig. IV–2).

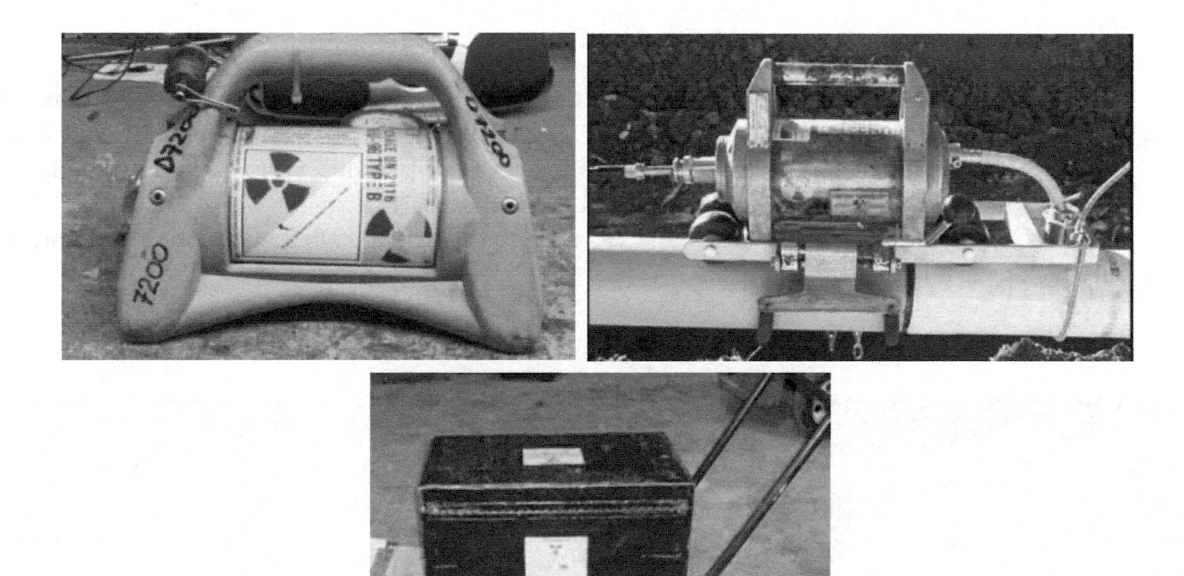

FIG. IV–1. Delta Sigma 880 camera, GammaMat M6 camera and a Tech Ops 660 (Type B(U)-96 transport package/ container).

FIG. IV–2. Licensed storage facilities at the user's premises.

IV–3. REGULATIONS FOR SAFETY AND SECURITY

Safety

— The Canadian Nuclear Safety Commission (CNSC) Act and regulations;
— Nuclear substances and devices regulations;
— Radiation protection regulations;
— Transport and packaging regulations;
— Security regulations;
— Transport Canada regulations.

Security

Various security regulations exist, including Canadian Security REGDOC-2.12.3: Security of Nuclear Substances: Sealed Sources and Category I, II and III Nuclear Material, Version 2. Part A of REGDOC-2.12.3 sets out the minimum security measures that licensees must implement to prevent the loss, sabotage, illegal use, illegal possession or illegal removal of sealed sources.

Security requirements include primary and secondary containment, high security locks and alarms, and GPS tracking at all times (Fig. IV–3).

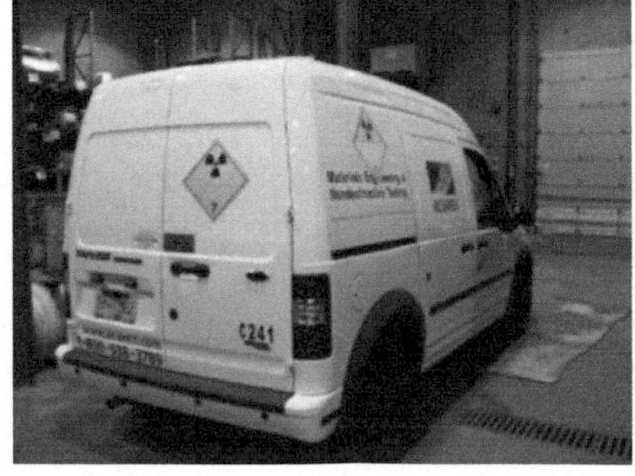

FIG. IV–3. Security process for transport of sources with DU as a shielding material (containment, locks and GPS tracking).

IV–4. EXPERIENCE REGARDING RETURN TO MANUFACTURER/SUPPLIER

Used or redundant radiographic exposure devices containing DU are usually returned to the manufacturer. Regulations allow users to transfer to another licensee if the device is listed on the user's licence.

In the past, old radiographic exposure devices were sent for long term storage at the Atomic Energy of Canada Limited nuclear facility at Chalk River in Ontario.

IV–5. EXPERIENCE REGARDING REUSE AND RECYCLING

Canada does not have any DU recycling services. Disposal may be through an authorized CNSC licensed facility with the device listed on licence or through transfer of a device to another CNSC licensed user. Devices containing DU must never be transferred to an unauthorized user or location.

IV–6. HAVE YOU IDENTIFIED ANY PROBLEMS OR NEED FOR ASSISTANCE FROM THE IAEA?

Tracking of DU should be specific to the container and not to the location, mass or quantity of the DU.

Annex V

MEMBER STATE EXPERIENCE: CHILE

V–1. TYPES OF DEVICES THAT CONTAIN DU

The majority of the volume of DU material is derived from used devices such as teletherapy units and industrial radiography equipment (Fig. V–1).

In addition, some DU collimators (from a teletherapy unit) were discovered in a scrap recycling company (Fig. V–2).

Table V–1 shows the inventory of DU material in Chile.

FIG. V–1. Radiography camera, collimator and a teletherapy unit containing DU as a shielding material.

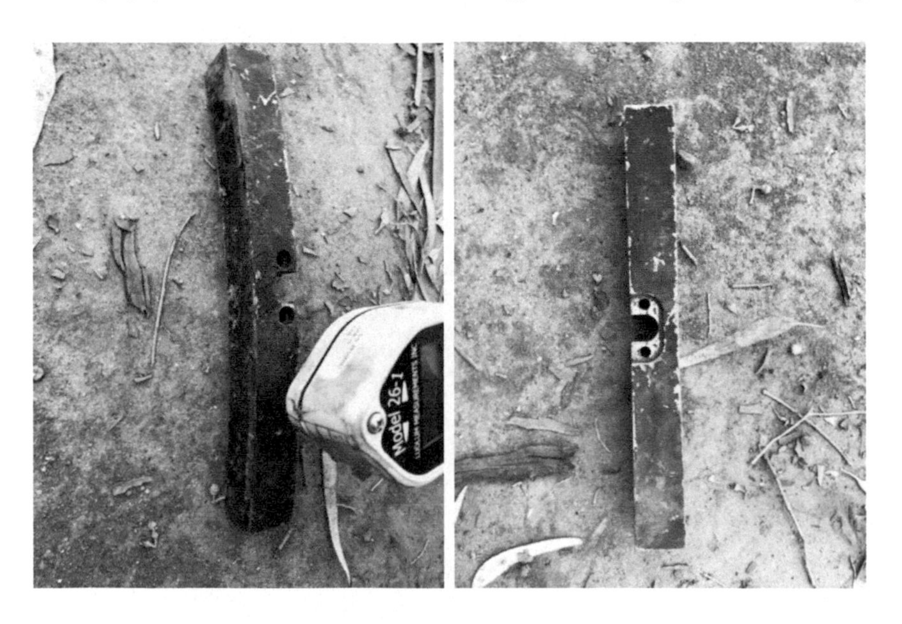

FIG. V–2. Examples of collimators found in scrap yards.

TABLE V–1. DU MATERIAL IN CHILE

Type	Number
Teletherapy unit	31
Industrial radiography equipment	110
Collimators	>30 pieces
Shielding	1

TABLE V–2. Devices containing DU stored at CCHEN facilities.

Type	CCHEN	Storage by user (or in use)
Teletherapy unit	29	3
Industrial radiography equipment	87	23
Collimators	>30 pieces	Unknown
Shielding	1	Unknown

V–2. STORAGE AND DISPOSAL FACILITIES FOR DU

All devices/containers with DU as a shielding material and containing DSRSs are kept in a CSF managed by the Chilean Nuclear Energy Commission (CCHEN).

All other devices/containers currently in use are under the control of the users — see Table V–2.

V–3. REGULATIONS FOR SAFETY AND SECURITY

Safety

Chile has four safety regulations:

- Law No. 18.302 of the 1982 Nuclear Safety Law defines nuclear substances such as nuclear fuel, with the exception of natural uranium and DU;
- Supreme Decree No. 133 of 1984 adopts regulation on the authorization of radioactive facilities or equipment generating ionizing radiation, personnel working in such facilities or operating such equipment and related activities;
- Supreme Decree No. 3 of 1984, Regulation on radiation protection and radioactive facilities, provides personal protection measures on radiological and radiation dose limits for occupationally exposed persons in order to prevent and avoid overexposure to ionizing radiation and its effects on health;
- Supreme Decree No. 12 of 1985, Regulations for safe transport of radioactive material, establishes the conditions to be met in the transport of radioactive materials, whether by land, water or air, while such radioactive materials cannot form an integral part of the means of transport.

In general, Chilean safety regulations do not make specific reference to DU.

Security

Chile has one Supreme Decree that addresses security issues:

- Supreme Decree No. 87 of 1985, Regulations on physical protection of facilities and nuclear materials, covers the evaluation, authorization and oversight of plans for the physical protection of nuclear facilities and nuclear materials under the responsibility of the Chilean Commission of Nuclear Energy, as the authority.
- On safeguards:
- 1967: Treaty of Tlatelolco (promulgated in 1974);
- 1995: Agreement between Chile and the IAEA is promulgated for the application of safeguards, and the NPT is promulgated;
- 2004: The additional protocol to the agreement with the IAEA is promulgated for the application of safeguards.

V–4. EXPERIENCE REGARDING RETURN TO MANUFACTURER/SUPPLIER

In 2013 the regulatory body informed users regarding expired transport certificate SPEC 2T. CCHEN has received DU shielding with ^{192}Ir DSRSs. The discovery of collimators found in scrap yards is described above.

The current practice in Chile is interim storage awaiting future conditioning and disposal.

V–5. EXPERIENCE REGARDING REUSE AND RECYCLING

Chile does not have any experience in recycling devices containing DU.

V–6. HAVE YOU IDENTIFIED ANY PROBLEMS OR NEED FOR ASSISTANCE FROM THE IAEA?

Chile is part of a collaborative project, “Enhancing nuclear security through the sustainable management of disused sealed radioactive sources in Latin America, Africa and the Pacific”, and the repatriation of Category 1–2 DSRSs is now a realistic option. However, this will leave more than 30 shielding containers empty, and the project does not consider the future management of the DU shielding.

Chile will need support to manage this DU shielding because its storage capabilities will be reduced once the DSRSs are repatriated.

Annex VI

MEMBER STATE EXPERIENCE: FRANCE

VI–1. TYPES OF DEVICES THAT CONTAIN DU

France has disused devices with DU shielding and pieces of DU used for:

— Medical applications: telegammatherapy head (Alcyon, Alcyon II, Theratron 780), comprising main head shielding, collimator and trimmers;
— Brachytherapy afterloader (Curietron series);
— Brachytherapy SRS storage and changer unit Curiestock for Curietron series.

Figures VI–1—15 show these devices.

DU shielding and DU pieces were also employed for industrial uses: industrial gammagraphy afterloaders (also called projectors or cameras); collimators and source storage/changing units; density, level and thickness gauges; and disused transportation containers for high activity material.

To a much lesser extent, DU shielding and DU pieces were used for R&D purposes, such as irradiation of samples in laboratories.

In terms of inventory of disused DU shielding, CEA and CIS Bio International store about 95 t of equipment containing 20 t of DU as of 2019. Total inventory of disused DU shielding in France was about 110 t as of 2016 (Andra National Inventory of Radioactive Waste and Radioactive Materials).

VI–2. STORAGE AND DISPOSAL FACILITIES FOR DU

There is presently no CSF for disused devices with DU shields. Some disused DU shields and devices are stored at user premises (e.g. disused medical devices at cancer treatment centres and hospitals), some at CEA centres, and some at the facilities of manufacturing companies. Andra has no facility for disused DU shielding.

Regarding DU disposal, Andra disposal facilities do not accept disused DU shielded devices at the CSFMA (LLW and ILW) and CIRES (VLLW) disposal facilities because these devices do not comply with their waste acceptance criteria. These facilities are licensed to dispose of waste contaminated with DU but not metallic DU blocks. Disposal of some small dimension blocks would call for additional safety assessment and modification of the licence application file.

FIG. VI–1. .Two views of an Alcyon telegammatherapy head before transportation Courtesy of L. Pillette-Cousin, France.

FIG. VI–2. The primary DU shield of an Alcyon head Courtesy of CIS Bio International, France.

FIG. VI–3. Two trimmer bars Courtesy of CIS Bio International, France.

FIG. VI–4. DU protection blocks supplied with a Theratron 780 unit. Courtesy of L. Pillette-Cousin, France.

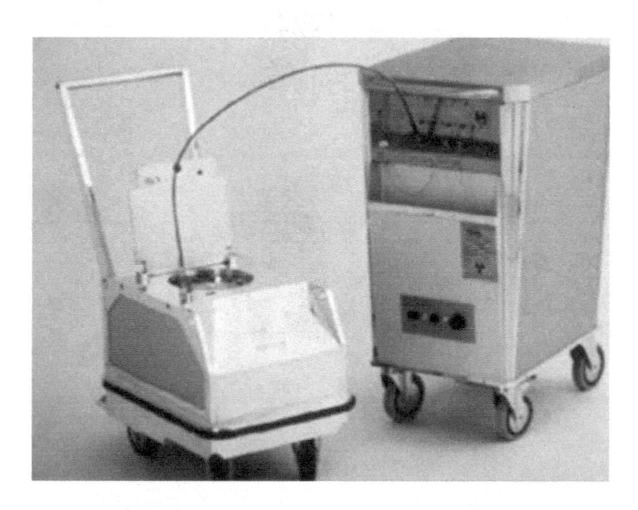

FIG. VI–5. A disused Curietron brachytherapy afterloader (back) and a Curiestock SRS storage and changing unit (front). Courtesy of CIS Bio International, France.

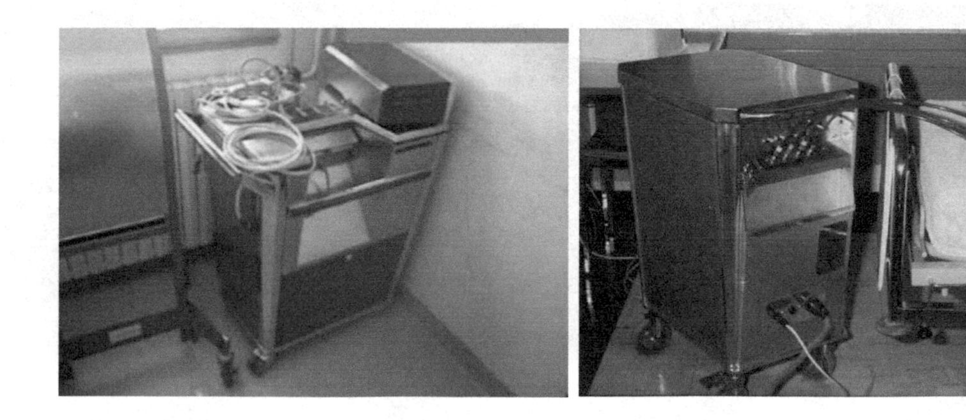

FIG. VI–6. Disused Curietron units. Courtesy of L. Pillette-Cousin, France.

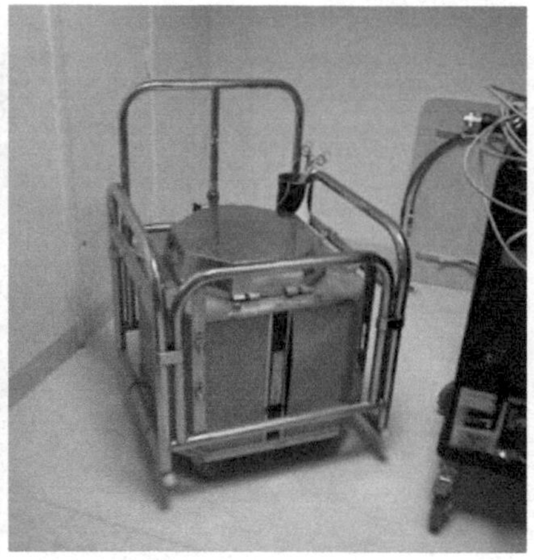

FIG. VI–7. Disused Curiestock units. Courtesy of CIS Bio International, France.

FIG. VI–8. A blood irradiator IBL437Courtesy of CIS Bio International, France.

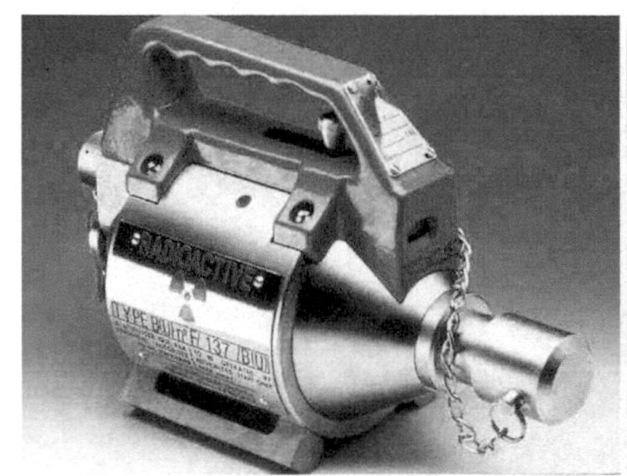

FIG. VI–9. (left) a portable gammagraphy projector GAM80 containing, when in use, an 192Ir SRS up to 3 TBq, (centre) presents a mobile gammagraphy projector GR50 containing, when in use, a 60Co SRS up to 1.85 TBq (right) shows a mobile gammagraphy projector GMA2500 containing, when in use, a 60Co SRS up to 7.4 TBq. Courtesy of Actemium NDT-P&S — Vinci Energies, France.

FIG. VI–10. A disused MAG 120 directional collimator for 192Ir SRSs. Courtesy of L. Pillette-Cousin, France.

FIG. VI–11. A disused directional (left) and panoramic (right) collimator for 192Ir SRSs. Courtesy of Actemium NDT-P&S — Vinci Energies, France.

FIG. VI–12. A directional collimator for 60Co SRSs. Courtesy of Actemium NDT-P&S — Vinci Energies, France.

FIG. VI–13. *The broken window of the disused directional collimator and the uranium block is visible. Courtesy of L. Pillette-Cousin, France.*

FIG. VI–14. *Disused SV17 transportation casks (left) and storage casks used by CIS Bio International (right). Courtesy of CEA, France.*

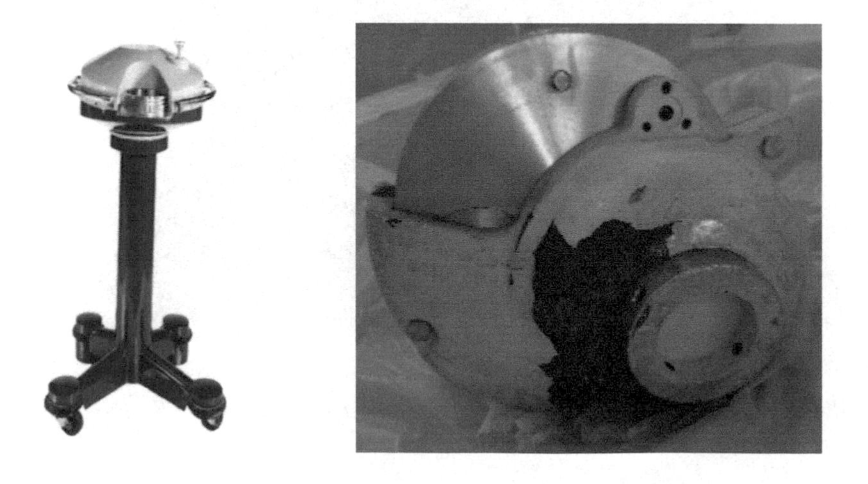

FIG. VI–15. *An Ammonite sample irradiator (left) and its DU shielding (right). Courtesy of CEA, France.*

VI–3. REGULATIONS FOR SAFETY AND SECURITY

Regulations are enforced regarding radiological safety, based on European directives, transportation of radioactive material (ADR), national safeguards and security dispositions, and industrial safety requirements. Indeed, facilities where disused DU shielding is stored (no DU shielding is processed, reused or recycled at present) are subject to several regulations:

- Basic nuclear facility regulations (Ministry of Ecological Transition);
- French and international safeguards (Ministry of Industry);
- Radiological safety regulation for workers, the public and the environment (Ministry of Health);
- Transport of radioactive material regulation (derived from IAEA SSR-6 and ADR) plus the Code of Defence.

Security of DU shielding is ensured according to French national safeguards requirements and regulations applicable to the protection of radioactive sources and radioactive material. Moreover, the Ministry for Labour has issued edicts on specific security requirements for gamma radiography equipment (e.g. for storage and transport conditions).

VI–4. EXPERIENCE REGARDING RETURN TO MANUFACTURER/SUPPLIER

Return to manufacturer is normally performed for disused gamma radiography equipment. It is not systematically the case for disused medical equipment. CEA accepts the return of equipment manufactured in the past if it can be proven that CEA was the manufacturer (certificates, pictures, etc.).

VI–5. EXPERIENCE REGARDING REUSE AND RECYCLING

Up to 1992, SICN, a front end fuel cycle company, was able to recycle some low amounts of DU shielding. Since 1992, no further DU recycling service has existed in France, and no recycling has occurred up to now with other Member States.

CEA and CIS Bio International cooperated for years through a public interest group called 'the groupement d'interet public (GIP) sources HA' to manage high activity disused radioactive sources. Thus, CEA manages an important part of DU shielding from radioactive source containers used both in and outside France.

With the French Ministry for Foreign Affairs and the IAEA, CEA is currently assessing a long term management solution for disused DU shielding in order to recover it, as well as DSRS.

At the same time, long term DU management solutions are being assessed by CEA, as summarized in Fig. VI–16.

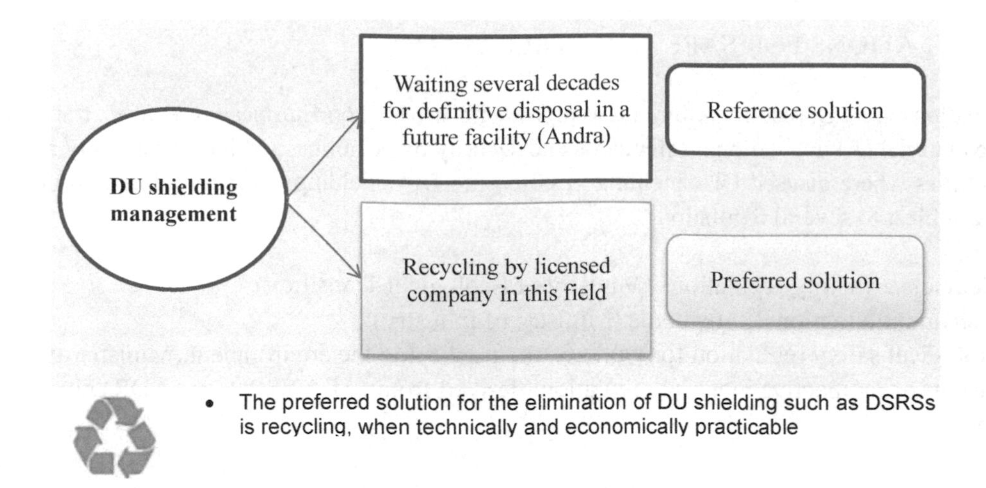

FIG. VI–16. Assessment of long term management solution for DU materials

The current conclusion of the CEA assessment is that the preferred solution for the elimination of disused DU shielding is recycling. It is at present a more expensive solution, even with all the uncertainties attached to cost evolutions for recycling, storage and disposal, but from the societal and environmental standpoints, recycling is better for future generations.

VI–6. HAVE YOU IDENTIFIED ANY PROBLEMS OR NEED FOR ASSISTANCE FROM THE IAEA?

No. France is happy to share its experience with other Member States in relation to the management of disused DU shielding. Member States having DU in disused medical or industrial devices should officially make a request to the French Ministry for Foreign Affairs to assist them with the repatriation of these DU shields to France.

Annex VII

MEMBER STATE EXPERIENCE: GEORGIA

VII–1. TYPES OF DEVICES THAT CONTAIN DU

In Georgia the nuclear and radiation sources registry contains DU containers in the following groups by their origin: legal containers (medicine and industry), historical (no longer used, from former USSR period), orphan and those from criminal activity.

Legal containers are used in medicine and industry, as well as for calibration (two ^{60}Co teletherapy devices with DU shielding) and non-destructive testing (NDT), mainly for pipelines (Delta-880 devices, TechOps, Model 660), standardization and metrology in Secondary Standard Laboratories (Isotope Technologies) — see Fig. VII–1.

Some historical transport containers (former USSR) were used by oncological hospitals (without responsibilities and agreement to 'send back to manufacturer' ^{60}Co and DU containers). Some historical DU containers were used for NDT in former USSR activities (^{137}CS, ^{192}Ir), and other DU containers were from an old former USSR (non-operational) scientific research reactor (for sources and irradiated samples).

More than 300 orphan sources were left in Georgia following the demise of the USSR.

Also left were DU containers used for ^{192}Ir sources (NDT use) and an abandoned radiotherapy device — see Fig. VII–2.

There is also an example of a DU container arising from criminal activity where the DU container was being sold on the black market — see Fig. VII–3.

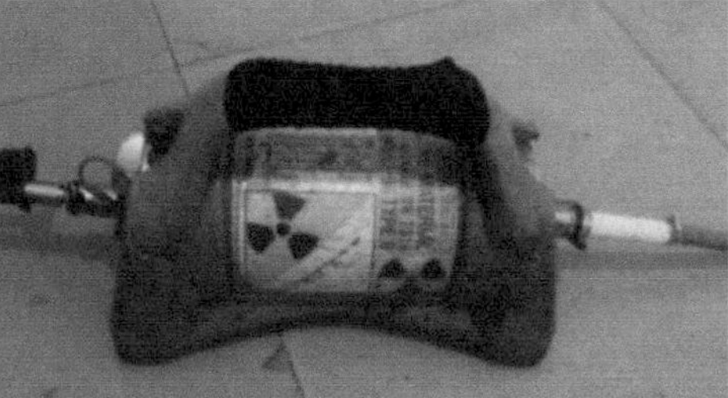

FIG. VII–1. Radiotherapy device, NDT gamma camera and device for standardization and metrology.

FIG. VII–2. Recovered orphan sources.

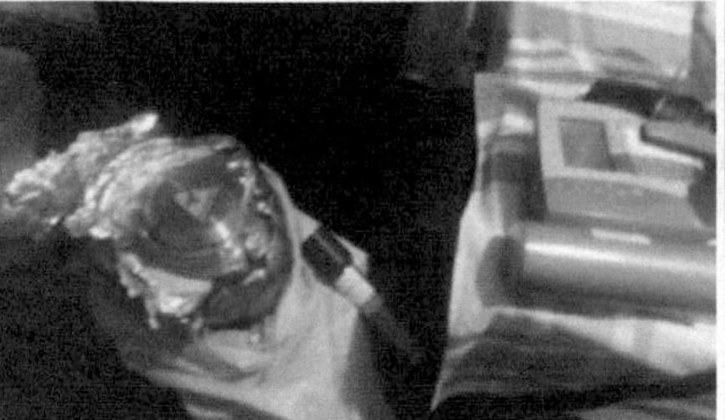
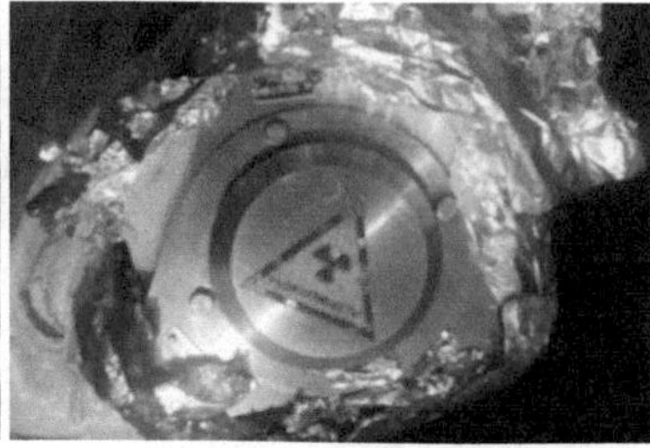

FIG. VII–3. Transport container containing DU shielding found stolen by criminal activity.

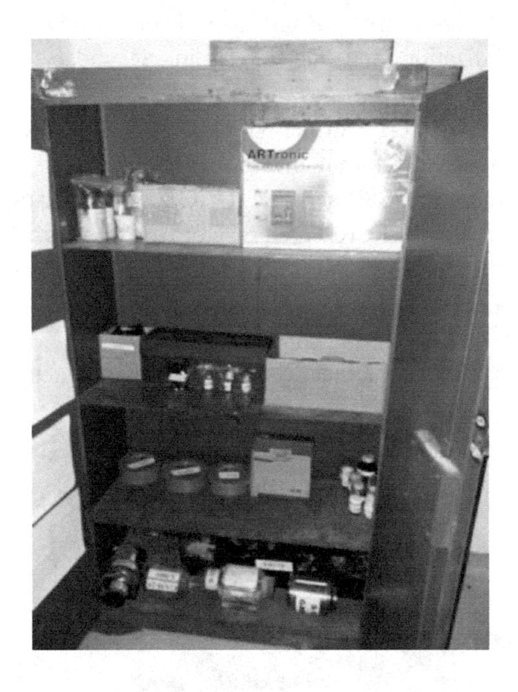

FIG. VII–4. DU containers in storage.

VII–2. STORAGE AND DISPOSAL FACILITIES FOR DU

All disused and found DU containers, as well as those from criminal activity, are kept in the nuclear and radiation waste management facilities in separate storage modules (Fig. VII–4). Some of these contain disused sources. A CSF (Fig. VII–4) has been in operation since 2007 to manage radioactive waste streams from the decommissioning of the IRT-M research reactor and radioactive sources from activities undertaken in Georgia. An older facility at Saakadze contains a substantial amount of radioactive waste and is considered to be a storage facility. A number of low activity disused radioactive sources are also kept at stores belonging to Georgia's Ministry of Defence.

Legal containers are kept at the licensee's premises until they can be sent back to the manufacturer.

VII–3. REGULATIONS FOR SAFETY AND SECURITY

Safety

Georgia does not have any special safety regulations for the management of DU containers. These are considered as sources of ionizing radiation, and all regulations related to sources also cover DU containers.

The corresponding Law of Georgia on Nuclear and Radiation Safety (chapter XI) states that all licensees shall present for import permission (sources, devices, associated equipment) a special agreement with the manufacturer that stipulates 'send back to manufacturer'.

Security

Georgia has no special security regulations for the management of DU containers, but they are considered as sources of ionizing radiation. DU containers are considered as nuclear material and are subject to the safeguard agreement with the IAEA and the Convention on the Physical Protection of Nuclear Material and its Amendment.

Nuclear material management issues are considered in the Law of Georgia on Nuclear and Radiation Safety[1].

Bylaws cover the physical (protection) security of nuclear and radiation facilities, radioactive sources and waste, and the procedure for carrying out activities connected to nuclear non-proliferation safeguards.

VII–4. EXPERIENCE REGARDING RETURN TO MANUFACTURER/SUPPLIER

Georgia has no radiation waste disposal facility yet, and disposal of disused DU containers is not being considered. All DU is kept in interim storage awaiting a final decision regarding disposal.

VII–5. EXPERIENCE REGARDING REUSE AND RECYCLING

Georgia does not have experience or future plans to recycle DU containers within or outside of Georgia. Currently there is a small inventory of disused DU containers (eight radiotherapy heads and nine or ten NDT containers), which are kept in interim storage. To create an organization for recycling of disused DU containers does not have an economical basis. All new imported DU containers will be sent back to the manufacturer and the regulatory body does not expect to see an increase in the use of DU as shielding in the future.

VII–6. HAVE YOU IDENTIFIED ANY PROBLEMS OR NEED FOR ASSISTANCE FROM THE IAEA?

The regulatory body has not identified any significant issues with DU containers and does not require assistance from IAEA at this point of time. However, with the possible establishment of a future national waste strategy and a national waste disposal facility, the issue of possible disposal of DU containers will need to be considered, including safeguards declaration issues — termination, exemption and other.

---

[1] www.anrs.gov.ge

Good practice in Georgia

From 2004, ANRS implemented ARIS software for the upkeep and maintenance of a register of sources;

— Nuclear material (including DU containers) is considered as a source of ionizing radiation and is recorded as ordinary sources (e.g. ^{137}Cs, ^{60}Co);
— Nuclear material is stored in separate modules and labelled with regulatory IDs and batch names in a dedicated interim waste storage facility;
— Nuclear material (DU containers) is also registered in a different database, NUCMAT, developed with USNRC support. The program has a reporting system for elaboration of the inventory change reports (ICRs), the material balance reports (MBRs) and the physical inventory listing (PILs).

Annex VIII

MEMBER STATE EXPERIENCE: HUNGARY

VIII–1. TYPES OF DEVICES THAT CONTAIN DU

Hungary has very limited inventories of DU and does not produce any DU. The DU inventory comprises predominantly industrial gamma (camera) radiography devices containing DU shielding. These include the following radiography cameras: EXERTUS DUAL 120, GammaMat M10, M18, SE, TI, TSI, TI-F, TSI 3/1 (Fig. VIII–1).

B(U) type foreign origin transport containers with DU shielding are also used. These include the UKT1B-50 and UK-12S for transporting high activity sealed sources.

There are several hospitals where DU is in use as an inner shielding layer in teletherapy devices, but there are only a few of these devices, as the usage of linac equipment is the preferred choice compared with teletherapy machines using ^{60}Co sources in Hungarian medical practice.

VIII–2. STORAGE AND DISPOSAL FACILITIES FOR DU

There are only a few companies that have a licence to store disused devices with DU shielding at their premises (mainly disused radiography devices). The largest inventory of DU is stored at the centralized disposal facility — the Radioactive Waste Treatment and Disposal Facility (for institutional radioactive waste) located at Püspökszilágy, 50 km north-east of Budapest (Fig. VIII–2).

An annual volume of 5 to 10 m^3 LLW and ILW, about 300 disused radioactive sources and 3000 radiation sources from smoke detectors (generated by smaller radioactive waste producers, such as hospitals, laboratories and industrial companies) are received at this facility. The most frequently disused isotopes are ^{60}Co, ^{137}Cs, ^{90}Sr and ^{3}H.

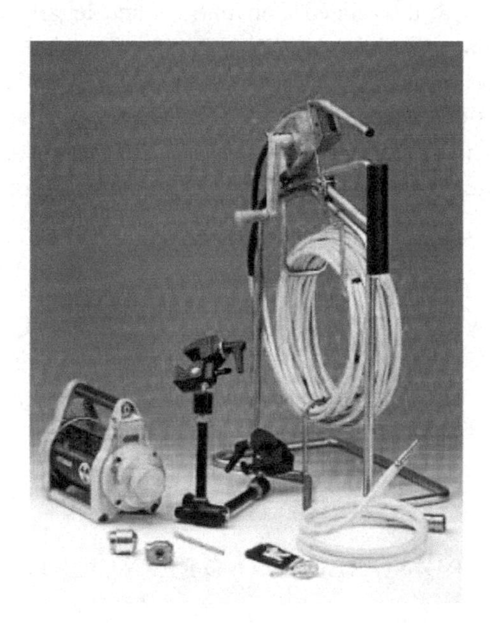

FIG. VIII–1. Industrial radiography cameras containing DU as a shielding material.

FIG. VIII–2. Institutional LLW and ILW is stored at the Radioactive Waste Treatment and Disposal Facility (including storage of DU and DSRSs).

VIII–3. REGULATIONS FOR SAFETY AND SECURITY

Safety

Radiation safety becomes a concern only when radioactive sources are inside a device containing DU requiring appropriate safety management protocols. (The basis of the Hungarian radiation safety legislation is the implementation of 2013/59/Euratom Basic Safety Standards in national legal acts.)

For transport safety, Hungary follows the provisions of the mode specific transport regulations for transporting dangerous goods (e.g. the International Carriage of Dangerous Goods by Road (ADR), Regulations concerning the International Carriage of Dangerous Goods by Rail (RID), The European Agreement concerning the International Carriage of Dangerous Goods by Inland Waterways (ADN)) and the requirements of IAEA Regulations for the Safe Transport of Radioactive Material, SSR-6 (Rev. 1).

Security

In Hungary, a comprehensive nuclear security regulatory regime has been established. The requirements are based on the recommendations of IAEA Nuclear Security Series Nos 9, 13, 14, 15 and 20, and are in accordance with the related Conventions and legally non-binding international legal instruments. Applying the nuclear material categorization recommended by IAEA, DU must be protected during use, storage and transport, at least in accordance with cost effective and appropriate management practice. In the Hungarian physical protection legislation, this means the application of D level security measures, which can be found in Government Decree 190/2011 (IX.19.) Korm on physical protection requirements for various applications of atomic energy and the corresponding system of licensing, reporting and inspection.

VIII–4. EXPERIENCE REGARDING RETURN TO MANUFACTURER/SUPPLIER

Some companies temporarily store the disused DU containers and devices for possible future economic benefit. Returning of DU devices to the originating manufacturer is not a common practice in Hungary.

VIII–5. EXPERIENCE REGARDING REUSE AND RECYCLING

Hungary has no experience in recycling devices containing DU.

VIII–6. HAVE YOU IDENTIFIED ANY PROBLEMS OR NEED FOR ASSISTANCE FROM THE IAEA?

In Hungary there are a significant number of radiography devices with DU shielding (GammaMat types) where package certificates have expired or will be expiring without a prolongation possibility. If these devices are not permitted for any further ongoing use as B(U) type package containers, they will become disused DU-containing devices.

Annex IX

MEMBER STATE EXPERIENCE: INDONESIA

IX–1. TYPES OF DEVICES THAT CONTAIN DU

The DU inventory (as shielding material) in Indonesia is contained within radioactive source transport containers, camera radiography projectors and collimators (Fig. IX–1).

The inventory and types of radiography cameras are included in Table IX–1 below (data from 2010).

IX–2. STORAGE AND DISPOSAL FACILITIES FOR DU

In Indonesia, all radioactive waste is stored in a CSF at the Centre for Radioactive Waste Technology — National Nuclear Energy Agency (NNEA) (Figs IX–2 and IX–3).

TABLE IX–1. Inventory of radiography cameras in Indonesia.

No.	Product	Radioactive source	Activity	Total
1	Amertest	Ir-192, Se-75	80–100	14
2	GammaMat	Ir-192, Se-75	80–100	97
3	Tech Ops 660	Ir-192, Se-75	80–100	298
4	Tech Ops 680	Co-60	100	2
5	Sentinel	Se-75	50–150	100
6	Iridia	Ir-192	80–100	2
7	SCAR	Ir-192	20	3
				516

FIG. IX–1. DU contained in a radiography camera and a radioactive source transport container.

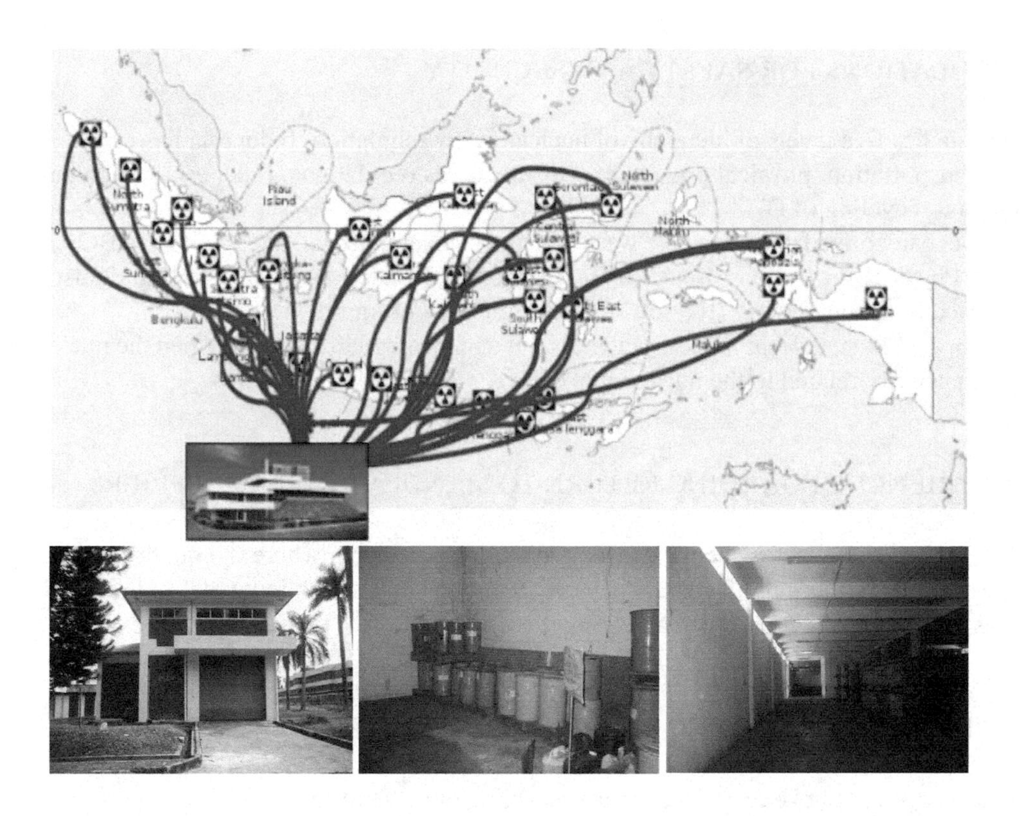

FIG. IX–2. The Centre for Radioactive Waste Technology — NNEA provides management services for radioactive waste from all Indonesian areas, including DU shielding material.

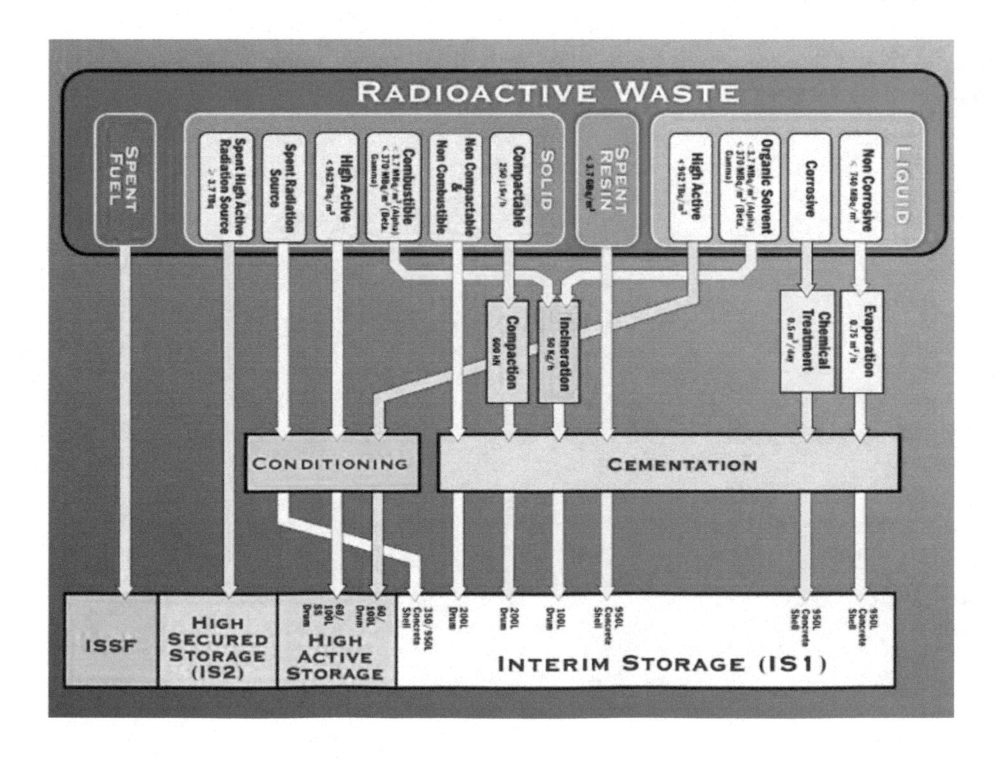

FIG. IX–3. Radioactive waste management at NNEA (Indonesia), including integrated waste storage facilities (also for DU) (courtesy of Nuclear Energy Regulator Agency, Indonesia)

IX–3. REGULATIONS FOR SAFETY AND SECURITY

Indonesia has five levels of hierarchy of nuclear safety regulation. Indonesia has regulations related to DU, i.e. transportation, physical protection and safeguards regulations. Indonesia has no policy on the enrichment and recycling of DU.

DU is not categorized as radioactive waste but as a nuclear material, as explained in BCR no. 1/2009: Provision of Physical Protection of Nuclear Installation and Materials. BCR no. 1/2009 also notes that DU is classified as group number III ($\geq$ 500 kg) and group IV (1 kg $<$ U/Th $<$ 500 kg).

BCR no. 4/2011 states that nuclear materials fall under safeguards control when the nuclear material is DU used in activity related to the nuclear fuel cycle.

IX–4. EXPERIENCE REGARDING RETURN TO MANUFACTURER/SUPPLIER

Indonesia has some experience related to radiographic cameras where the quality certificates were not being extended by the manufacturer. As such, Tech Ops 660 series industrial radiographic cameras used both as radiographic cameras and as transport containers cannot be used any more.

IX–5. EXPERIENCE REGARDING REUSE AND RECYCLING

Indonesia does not have experience in the recycling or reuse of DU.

IX–6. HAVE YOU IDENTIFIED ANY PROBLEMS OR NEED FOR ASSISTANCE FROM THE IAEA?

Indonesia would like to benefit from some form of assistance in the most cost effective process for reusing and recycling DU.

Annex X

MEMBER STATE EXPERIENCE: ISLAMIC REPUBLIC OF IRAN

X–1. TYPES OF DEVICES THAT CONTAIN DU

Devices containing DU include portable industrial gamma projectors, teletherapy heads, source containers/devices and logging tools (Fig. X–1).

The specification of all the radiation sources and devices (including those that contain DU as shielding) are logged in a regulatory database and information bank, including owner, serial number, manufacturing company name, model, application, weight (for DU), categorization (for sources), activity (for sources) and the production date.

X–2. STORAGE AND DISPOSAL FACILITIES FOR DU

Portable industrial gamma projectors

After 2015, the Iranian Nuclear Regulatory Authority (INRA) decided that some models and types of portable industrial gamma projectors should be taken out of further service, as the models were no longer compliant with the requirements of ISO-3999 standards and the certificate of approval of type B package transport had expired.

As a result, owners of specific model gamma projectors must now deliver their projectors to the waste management organization within a specific time period (deadline).

The Iran Radioactive Waste Organization (IRWA) is designated by the Atomic Energy Organization of Iran as the central waste management organization.

Teletherapy heads

Almost all the teletherapy heads (containing high activity ^{60}Co sources) have been sent to the waste management organization for storage.

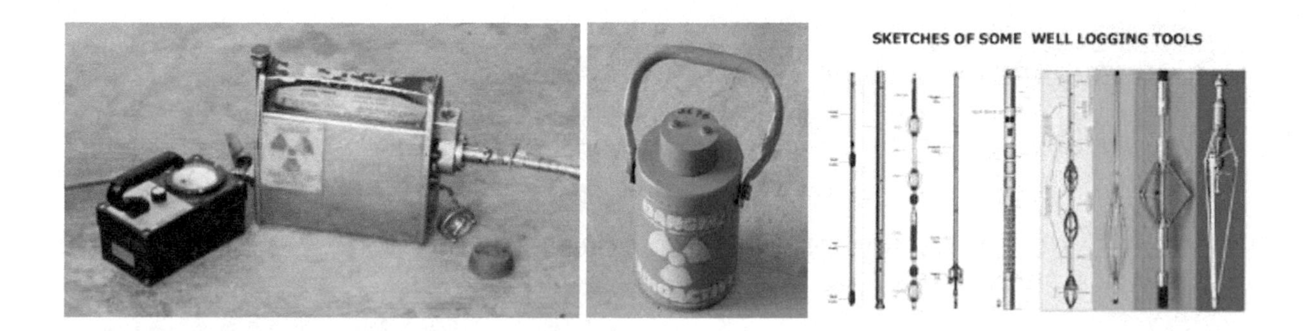

FIG. X–1. Portable gamma devices, transport containers and well logging equipment that contains DU as a shielding material.

Source containers

Some source containers that have DU for shielding are now stored at the waste management organization awaiting dismantling.

Inventory

Approximately six barrels containing conditioned disused sealed sources, 25 teletherapy heads (Theratronics, Mobalteron and Picker type) with about 1000 Ci activity, 400 ^{60}Co sources, 200 ^{137}Cs sources, 60 Am-Be, Ra-Be and ^{252}Cf sources and 3000 ^{192}Ir sources are stored in the IRWA CSF.

Well logging tools

All well logging tools that contain DU have been returned to the supplier as required.

X–3. REGULATIONS FOR SAFETY AND SECURITY

Safety

Safety regulations include:

— The Radiation Protection Act of Iran;
— The Basic Radiation Safety Standards (BRSS);
— The Decree of Radiation Protection Act of Iran;
— Requirements and codes of practice for each application.

Security

The Islamic Republic of Iran has some general regulations regarding security but not specific ones. All regulations are under the safeguards agreement with the IAEA.

X–4. EXPERIENCE REGARDING RETURN TO MANUFACTURER/SUPPLIER

Some well logging equipment (containing DU) has been returned to the original supplier. Experience has been gathered with the transport of used devices and DSRSs (containing DU) being packaged and transported to central storage at the IRWA site.

X–5. EXPERIENCE REGARDING REUSE AND RECYCLING

Not for recycling — only return to the supplier where possible.

X–6. HAVE YOU IDENTIFIED ANY PROBLEMS OR NEED FOR ASSISTANCE FROM THE IAEA?

Assistance is needed to appropriately manage the DU within containers and devices currently stored at the central waste management organization storage facility.

In the case of portable industrial gamma projectors and source containers that contain DU, it is preferred to return them to the supplier or other company for reuse and/or a swap for new shielded devices.

In the case of teletherapy heads, the Islamic Republic of Iran would like to receive assistance with packaging and transporting them to the supplier or another company for recycling.

Annex XI

MEMBER STATE EXPERIENCE: JORDAN

XI–1. TYPES OF DEVICES THAT CONTAIN DU

Table XI–1 shows the inventory for disused teletherapy machines stored at the national CSF with DU as a shielding material. Figure XI–1 shows images of a teletherapy machine and dismantled heads.

XI–2. STORAGE AND DISPOSAL FACILITIES FOR DU

The teletherapy heads have been dismantled and transported from the user's premises to the national CSF located in the Jordan Atomic Energy Commission (JAEC) headquarters for storage (Fig. XI–2).

There are four storage rooms in the CSF, which store characterized and uncharacterized radioactive waste and DSRSs in their original shield packages or containers, empty cemented 200 L steel drums to be used for conditioning radioactive waste and for storing of neutron sources, and a dedicated storage room that contains conditioned radioactive waste/DSRSs in cement-lined 200 L drums, as well as some DU and a teletherapy head source within its original shield container.

Most of the DU stored at the CSF is contained in teletherapy medical devices as a shielding material for ^{60}Co (Category 1–2) sources. The national plan is to return the sources back to the supplier (in Canada), while other devices containing smaller quantities of DU that were used as counterweights for aircraft are now stored at the CSF and have no future use.

TABLE XI–1. Inventory of disused teletherapy machines with DU as a shielding material, stored at the national CSF.

Manufacturer	Model	Radionuclide
THERATRON 780c	S-5102	Co-60
PHOENIX	S-4631	Co-60
SIEMENS	Gammatron S80	Co-60

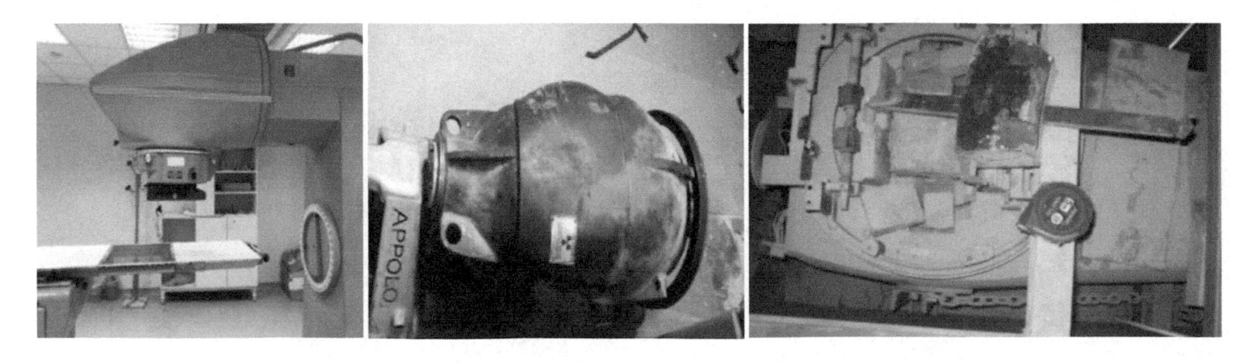

FIG. XI–1. Teletherapy machine and dismantled teletherapy heads.

FIG. XI–2. CSF located at the JAEC site.

XI–3. REGULATIONS FOR SAFETY AND SECURITY

Safety

DU is considered as radioactive waste and complies with the nuclear safeguards agreement (IAEA). The nuclear regulatory body for Jordan must submit an annual report to the IAEA advising of the inventory of DU existing in Jordan, including all technical specifications of devices containing DU.

Some applicable safety regulations/laws include:

— Law No. 43, Radiation Protection and Nuclear Safety and Security, Amman, Jordan 2007;
— EMRC, Radiological Protection System (Regulation No. 108), Amman, Jordan, 2015;
— EMRC, Instructions on Documents and Data to be Attached and Submitted along with the Filled Transport Application Form for Transport of Radioactive Materials, 2015;
— EMRC, Instructions for Specific Basis and Standards for Classification of the Type and Category of each RM Transport Package, 2015;
— EMRC, Instruction on Licensing Requirements for Issuing Transport of Radioactive Material, 2015;
— EMRC, Regulation No. (32) on Radioactive Material Transportation, Amman, Jordan, 2016.

Security

The regulatory body has implemented comprehensive safety and security regulations and instructions on radioactive waste and SNF management. There is always a security escort during the transport of devices containing DU as a shielding material, as well as a sophisticated security system at the CSF.

XI–4. EXPERIENCE REGARDING RETURN TO MANUFACTURER/SUPPLIER

CSF staff have first-hand experience in the field of storage and transport of radioactive materials and radiation sources.

CSF staff in 2009 collected and transported (for storage in the CSF) a radiotherapy head with DU as a shielding material from a government hospital.

CSF staff, with assistance from IAEA experts, in 2016 dismantled and transported two teletherapy heads with DU as a shielding material from a government hospital to the CSF.

XI–5. EXPERIENCE REGARDING REUSE AND RECYCLING

Jordan has no practical experience in recycling any type of device containing DU.

Jordan's national policy, entitled National Policy for Radioactive Waste and Spent Nuclear Fuel Management, promotes the option to repatriate sources and DU back to the supplier (country of origin) for recycling.

XI–6. HAVE YOU IDENTIFIED ANY PROBLEMS OR NEED FOR ASSISTANCE FROM THE IAEA?

Jordan needs IAEA technical assistance for the return of two medical teletherapy units (^{60}Co) of Category 1–2 to their origin country.

Jordan also needs IAEA technical assistance in drafting and implementing specific regulations and instructions pertaining to safe management options for DU.

Annex XII

MEMBER STATE EXPERIENCE: KENYA

XII–1. TYPES OF DEVICES THAT CONTAIN DU

Devices containing DU are predominantly ^{192}Ir gamma cameras and, previously, a number of ^{60}Co teletherapy heads (Fig. XII–1). There may also be some ^{137}Cs LDR sources (yet to be confirmed).

XII–2. STORAGE AND DISPOSAL FACILITIES FOR DU

All active sources and devices in use are kept at the user's premises. The DU with DSRSs is collected and kept at the storage facility of the regulator, the Radiation Protection Board (RPB) (Fig. XII–2). RPB members include the director of medical services (as chair), the chief radiologist and one medical specialist.

The RPB oversees matters of radiation safety, security of radioactive materials, control of contamination of food and the environment, and maintains a register of radiation facilities, radiation workers, suppliers of radioactive devices and materials. At present there is no national policy on the

FIG. XII–1. Iridium-192 gamma cameras.

FIG. XII–2. RPB office and facility and DSRS storage.

management of radioactive waste. Radioactive waste is conditioned and stored at a location operated by the Materials and Testing Research Laboratory. DSRSs are also stored by the RPB.

XII–3. REGULATIONS FOR SAFETY AND SECURITY

The Radiation Protection Act CAP 243 Laws of Kenya of 1986 addresses all matters relating to radiation safety.

There are no specific regulations for the management of DU as a waste material. DU material is treated in the same manner as any other (non-nuclear) radioactive material. The return to manufacturer option has been used in the past in Kenya.

Kenya is aiming to draft appropriate regulations once the Nuclear Regulatory Bill, 2018 becomes law. The bill has undergone a first reading phase and is now committed to the Departmental Committee on Energy. The new bill will include nuclear safety, security and safeguards.

— Storage: one DU source (about 15 kg);
— Transport: the same regulations and conditions are in place as for all radioactive material;
— Disposal: currently assessing a near surface repository disposal facility, to include DU materials.

XII–4. EXPERIENCE REGARDING RETURN TO MANUFACTURER/SUPPLIER

— Return to manufacturer: this is a preferred option and an acceptable method whenever possible to do so;

XII–5. EXPERIENCE REGARDING REUSE AND RECYCLING

None as yet, as the current DU inventory is small and the recycling option expensive.

XII–6. HAVE YOU IDENTIFIED ANY PROBLEMS OR NEED FOR ASSISTANCE FROM THE IAEA?

There is a need for assistance in a national training programme on how to process DU for interim storage and final disposal. Kenya has a large facility that can be utilized for such regional training.

Alternatively, assistance in the collection and export of DU to countries or facilities that can recycle the same would be welcomed by Kenya.

Annex XIII

MEMBER STATE EXPERIENCE: MADAGASCAR

XIII–1. TYPES OF DEVICES THAT CONTAIN DU

In Madagascar, we have devices containing DU including:

— 1 telegammatherapy device from China, which is no longer used. This device is stored at the hospital CHU HJRA. See Fig. XIII–1;
— 3 industrial radiography cameras. See Fig. XIII–2;
— 1 seed irradiator with 60Co sources. See Fig. XIII–3.

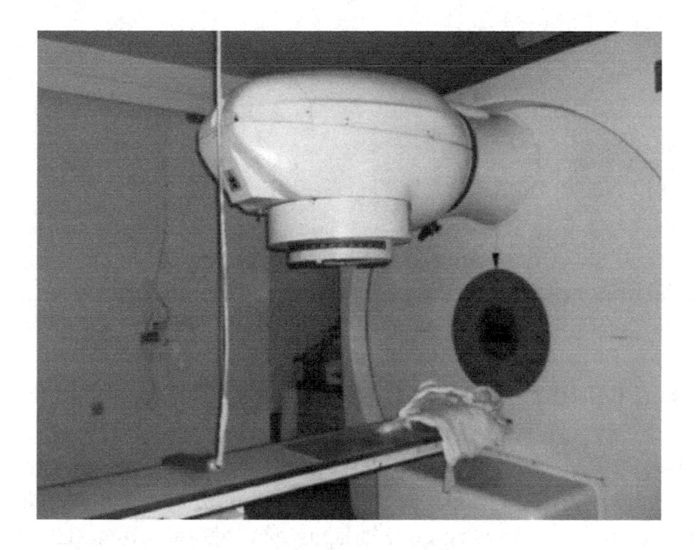

FIG. XIII–1. Telegammatherapy device from China, which is no longer used.

FIG. XIII–2. Industrial radiography cameras.

FIG. XIII–3. Seed irradiator with ^{60}Co sources.

XIII–2. STORAGE AND DISPOSAL FACILITIES FOR DU

Currently, there is no CSF in the country. The sources are stored in the user's premises: the hospital (Fig. XIII–4, left) or the laboratory for radioisotopes (Fig. XIII–4, centre and right).

XIII–3. REGULATIONS FOR SAFETY AND SECURITY

Madagascar Law No. 97-041 (2 January 1998) on protection against the danger of ionizing radiation and the management of radioactive waste in Madagascar exists. Additionally, Decree No. 2002-1274 (7 October 2002) on the management of radioactive waste in Madagascar is also in effect. However, DU is not included in this decree. There is a plan to update the decree to include DU.

XIII–4. EXPERIENCE REGARDING RETURN TO MANUFACTURER/SUPPLIER

Some experiences have concerned the dismantling of a 60Co teletherapy unit, prior to repatriating the source to France (Figs XIII–5 and XIII–6). DU metal parts were removed from the head of the ^{60}Co unit. Unfortunately, these DU shields were stolen from the hospital's storage site.

FIG. XIII–4. Sources are stored at the hospital (left) or laboratories (centre and right).

FIG. XIII–5. Type B(U) package with the teletherapy unit fixed on the vehicle.

FIG. XIII–6. Overpacking the teletherapy unit into the transport container.

DU metal parts were removed from the head of the ^{60}Co unit according to the following steps:

A. Securing the various DU pieces and elements.

— Packaging of each DU piece containing:
  - A pair of cotton gloves;
  - A pair of latex gloves;
  - A wooden crate whose interior is covered with a lead sheet 0.2 mm thick.

— Closing the crate with spikes;

— Setting up the crate in a thick red vinyl bag to provide protection.

B. Securing the DU collimator.

The disassembled parts are successively packaged:

— Four DU attenuator bars;

— Four DU primary bars;
— One DU attenuator disc.

XIII–5. EXPERIENCE REGARDING REUSE AND RECYCLING

We do not have experience recycling DU.

XIII–6. HAVE YOU IDENTIFIED ANY PROBLEMS OR NEED FOR ASSISTANCE FROM THE IAEA?

No response given.

Annex XIV

MEMBER STATE EXPERIENCE: MALAYSIA

XIV–1. TYPES OF DEVICES THAT CONTAIN DU

These include gamma projectors, teletherapy heads and various industrial gauges, including density gauges (Figs XIV–1 and XIV–2).

XIV–2. STORAGE AND DISPOSAL FACILITIES FOR DU

The users manage and store the sources/DU shielding on a temporary basis before sending them to a CSF. Malaysia has a national storage facility (including DU material) located at the site of the Malaysian Nuclear Agency — Nuclear Malaysia located near Bangi, Selangor (Fig. XIV–3).

FIG. XIV–1. Collimator (left) and DU shielding material (right) from a teletherapy head.

FIG. XIV–2. Container with DU shielding from an industrial application (left) and a gamma projector (right).

FIG. XIV–3. Centralized radioactive waste management facility at the Malaysian Nuclear Agency.

XIV–3. REGULATIONS FOR SAFETY AND SECURITY

The gauges and containers (containing DU shielding) are managed in the same manner as all of the other radioactive waste, in accordance with the Radiation Protection (Radioactive Waste Management) Regulation 2011.

There are no specific security regulations.

The Malaysian Nuclear Regulator (AELB) exercises control and oversight over the production, application and usage of radioactive materials and nuclear technology. The requirements from AELB are that:

- Users of radioactive materials are responsible for the waste;
- Users are required to be registered and licensed by AELB;
- DSRSs and DU shielding material is appropriately stored or returned to supplier;
- Users without infrastructure and expertise can request third party organization for services that are licensed/recognized by AELB.

The licence holders are responsible for the safety and physical protection of the radioactive waste, including DU materials.

XIV–4. EXPERIENCE REGARDING RETURN TO MANUFACTURER/SUPPLIER

The gauges and containers containing DU are stored in a CSF (Fig. XIV–3). The empty gauges and containers are transported to the CSF as low specific activity material.

XIV–5. EXPERIENCE REGARDING REUSE AND RECYCLING

There is no experience in Malaysia in reuse or recycling in this area.

XIV–6. HAVE YOU IDENTIFIED ANY PROBLEMS OR NEED FOR ASSISTANCE FROM THE IAEA?

Yes. How to better identify devices and/or containers that have DU as a shielding material.

Annex XV

MEMBER STATE EXPERIENCE: PAKISTAN

XV–1. TYPES OF DEVICES THAT CONTAIN DU

Teletherapy heads (^{60}Co) and medical isotope product transport containers for ^{99}Mo (Fig. XV–1), ^{131}I and ^{177}Lu.

XV–2. STORAGE AND DISPOSAL FACILITIES FOR DU

Currently, there are two licensed interim storage facilities:

— PINSTECH for the capital territory and northern part of Pakistan;
— KANUPP for the southern part of Pakistan.

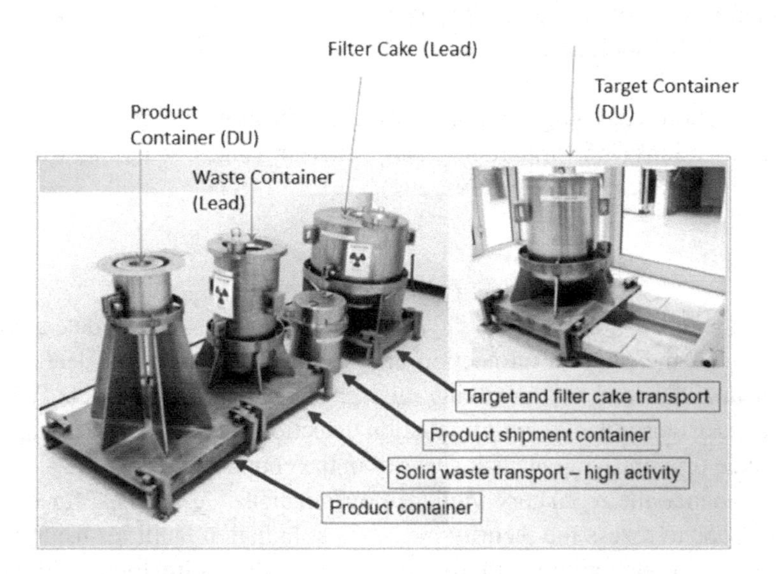

FIG. XV–1. Certified transport containers containing DU shielding used to ship ^{99}Mo medical isotope and other irradiated targets.

FIG. XV–2. PINSTECH interim CSF for radioactive waste, DSRSs and DU shielding material.

The interim storage facility at the Pakistan Atomic Energy Commission (PAEC)-owned Pakistan Institute of Nuclear Science and Technology — PINSTECH in Islamabad (Fig. XV–2) is one of two designated sites for the receipt and management of radioactive waste, including DSRSs generated from the operation of various radiation facilities in Pakistan.

XV–3. REGULATIONS FOR SAFETY AND SECURITY

Safety

The Pakistan Nuclear Regularity Authority (PNRA) is responsible for nuclear material inventory management. PNRA regulates the safe transportation of radioactive material in Pakistan in accordance with the regulatory requirements prescribed in Regulations for the Safe Transport of Radioactive Material — PAK/916. The consignments of radioactive material have to comply with national and international requirements. The authorization of such shipments is granted on acceptance of relevant documents by PNRA. PNRA also issues certificates for transport packages designed and manufactured in Pakistan. A central database of transport packages and shipments is maintained at PNRA.

A large number of radiation facilities are operating in Pakistan. PNRA is obliged to ensure that appropriate regulatory control is in place for the safe operation of these facilities. The radiation facilities being monitored by PNRA include:

- Medical applications (radiology, radiotherapy and nuclear medicine);
- Industrial applications (industrial radiography, nuclear gauges, scanning, irradiation, etc.);
- Research (universities, research centres, etc.).

Security

PNRA has made efforts to strengthen physical protection measures at nuclear and radiation facilities. The main objective of the physical protection and security is to prevent, detect and respond to theft, sabotage or other unauthorized acts involving nuclear or radioactive material. This is provided through training, technical assistance, deploying of radiation detection equipment, enhancing regulatory vigilance and issuing guidance on improving nuclear security in the country.

In order to enhance the regulatory vigilance over radioactive sources, for example, PNRA carries out regular inspections to assess the security measures at radiation facilities using high activity radiation sources. This is done in collaboration with regional nuclear safety directorates. Physical security plans of some radiation facilities are also reviewed to ensure these plans are as per PNRA requirements. Under the IAEA–Pakistan Nuclear Security Cooperation Programme, the security of three PAEC nuclear medicine centres at Islamabad, Abbottabad and Peshawar have been upgraded under the regulatory oversight of PNRA. Physical security upgrade plans for another 12 medical centres are under way.

XV–4. EXPERIENCE REGARDING RETURN TO MANUFACTURER/SUPPLIER

The PNRA Regulations PAK/908, PAK/904, PAK/916 and PAK/915 address the complete life cycle of DSRSs and associated DU shielding material, i.e. from authorization to safe disposal, including the return to a supplier option.

Currently there is no radioactive waste permanent disposal facility in Pakistan and interim storage is the common management method.

XV–5. EXPERIENCE REGARDING REUSE AND RECYCLING

Yes — for ^{60}Co, ^{99}Mo, ^{131}I, ^{177}Lu medical isotopes and DU(B) type containers. Recycling of these containers with DU shielding is within Pakistan only.

Pakistan also has a reuse/recycle policy for radioactive material, as per PAK 915 (13):

- Reuse and recycling of materials shall be applied to the extent possible to keep the generation of radioactive waste to the minimum practicable;
- The licensee using radioactive material shall not:
  - (a) Dismantle any sealed source without prior approval of the Authority;
  - (b) Transfer the material without confirmation that the organization to which it is to be transferred has the necessary authorization to hold/use/recycle that material.

XV–6. HAVE YOU IDENTIFIED ANY PROBLEMS OR NEED FOR ASSISTANCE FROM THE IAEA?

From time to time the competent authority (PNRA) may request/need future assistance from the IAEA. A follow-up with PNRA is recommended following the release of the Nuclear Energy Series Technical Document to the Member States.

Annex XVI

MEMBER STATE EXPERIENCE: ROMANIA

XVI–1. TYPES OF DEVICES THAT CONTAIN DU

A variety of devices/containers (some with DU as a shielding material) are used in Romania (Fig. XVI–1). These include gamma detectors, transport devices/containers, industrial gauges, teletherapy heads and collimators. The total mass of DU in Romania as a shielding material is approximately 8000 kg.

XVI–2. STORAGE AND DISPOSAL FACILITIES FOR DU

There are two CSFs in Romania:

— National Institute for Physics and Nuclear Engineering — Horia Hubei (IFIN-HH) Bucharest — Magurele (Fig. XVI–2 (left));
— Institute for Nuclear Research (ICN)-Pitesti (Fig. XVI–2 (right)).

XVI–3. REGULATIONS FOR SAFETY AND SECURITY

Safety regulations follow the same protocol as for other radioactive waste (including DSRSs).

Regulatory control is carried out by the National Commission for Nuclear Activities Control (CNCAN) through controlled authorizations of nuclear material related activities and by applying the

FIG. XVI–1. Gamma detectors, transport containers and teletherapy devices containing DU for shielding purposes.

FIG. XVI–2. (Left) DU storage area at IFIN-HH (courtesy of IFIN-HH) and (right) ICN-Piteste.

Euratom safeguards system. All equipment containing DU as shielding is owned and used only through specific authorization obtained from CNCAN.

The current Law no. 111/1996 empowers CNCAN to issue mandatory regulations on nuclear safety, radiological protection, quality assurance, transport of nuclear and other radioactive materials, management of radioactive waste and SNF; on emergency preparedness and intervention in case of nuclear accident; and in the manufacturing of products and supply of services for nuclear installations.

The current Law no. 111/1996 empowers CNCAN to issue mandatory regulations on nuclear safeguards/non-proliferation of nuclear weapons and physical protection of nuclear facilities and materials.

All small holders have a nominated person responsible for the site-specific safeguards needs (with approval from CNCAN), along with a specific reporting protocol.

XVI–4. EXPERIENCE REGARDING RETURN TO MANUFACTURER/SUPPLIER

Yes, for long term storage under safeguards control, transport and return to manufacturer.

XVI–5. EXPERIENCE REGARDING REUSE AND RECYCLING

No. However, equipment containing DU as radiation shielding can be exported, imported and transferred internally and intra-community only after obtaining specific authorization from CNCAN.

XVI–6. HAVE YOU IDENTIFIED ANY PROBLEMS OR NEED FOR ASSISTANCE FROM THE IAEA?

Require further work in the following areas:

- Waste acceptance criteria for a final repository to accept DU as radioactive waste and removal of safeguards control;
- Romania/CNCAN needs to update its records/database to ensure all of the DU shielding material in Romania is accounted for;
- It is essential to build a database to ensure traceability of all nuclear material and generate appropriate safeguards reports.

Annex XVII

MEMBER STATE EXPERIENCE: SENEGAL

XVII–1. TYPES OF DEVICES THAT CONTAIN DU

- — Linac, brachytherapy, radiotherapy devices (Fig. XVII–1);
- — Industrial gamma radiography;
- — Gauges for measurements of thickness, density, etc. (Fig. XVII–1);
- — Blood irradiators;
- — Gamma cameras;
- — Empty containers.

XVII–2. STORAGE AND DISPOSAL FACILITIES FOR DU

Yes; some empty containers (with DU shielding) are kept in a small storage facility but not the rest.

XVII–3. REGULATIONS FOR SAFETY AND SECURITY

No separate regulations for safety. Safety is covered in the global safety sense for all radioactive material as follows:

- — Convention on Nuclear Safety, in force: 24/03/2009;
- — Support of the IAEA Code of Conduct and its supplementary Guidance on the Import and Export of RS in July 2010.

No major experiences in security, but comply with the following for security and safeguards purposes:

SECURITY

- — Convention on the Physical Protection of Nuclear Material: in force 03/12/2003, Ratified (03/11/2003);
- — Amendment to the Convention on the Physical Protection of Nuclear Material; (from 05/04/2017).

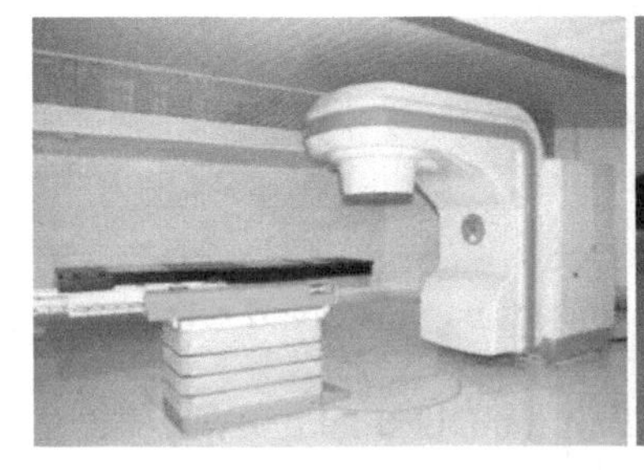
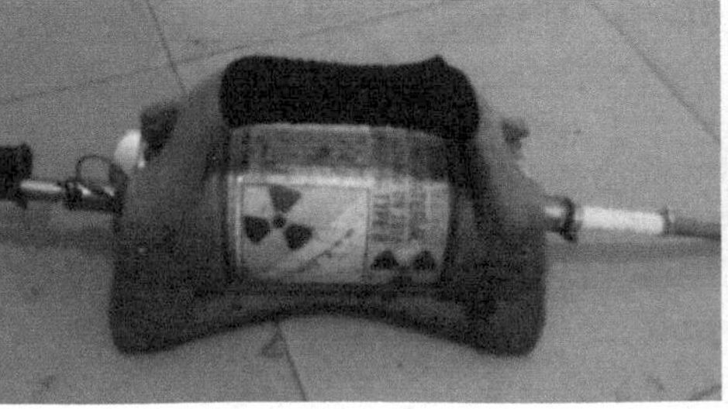

FIG. XVII–1. Radiotherapy head and industrial gauges containing DU shielding.

SAFEGUARDS

- NPT ratified (22/12/1970);
- Safeguards agreement in connection with the NPT (with protocol) (14 January 1980);
- Modified Small Quantities Protocol (SQPMod) signed 15 December 2006;
- Protocol(s) Additional to that Agreement approved in 2005 and signed in 2006; Ratified (from 05/04/2017).

XVII–4. EXPERIENCE REGARDING RETURN TO MANUFACTURER/SUPPLIER

Only with a single telegammatherapy unit containing ^{60}Co (re-exported) — see Fig. XVII–2.

XVII–5. EXPERIENCE REGARDING REUSE AND RECYCLING

None.

XVII–6. HAVE YOU IDENTIFIED ANY PROBLEMS OR NEED FOR ASSISTANCE FROM THE IAEA?

- To prepare a national inventory of DU shielding material for devices and transport containers in Senegal;
- To customize the inventory of such material in an appropriate database.

FIG. XVII–2. Repatriation of a teletherapy head containing DU and a ^{60}Co source.

Annex XVIII

MEMBER STATE EXPERIENCE: SLOVAK REPUBLIC

XVIII–1. TYPES OF DEVICES THAT CONTAIN DU

Transport containers, level gauges, flow gauges, density gauges, radiography devices, collimators and teletherapy heads (Fig. XVIII–1).

XVIII–2. STORAGE AND DISPOSAL FACILITIES FOR DU

Storage of DU occurs at:

— User/licensed premises;
— Central storage for nuclear materials (Fig. XVIII–2);
— Central storage for radioactive waste (in cases when the source is still inside a device or container);
— Return to manufacturer;
— Reuse by another licensed holder.

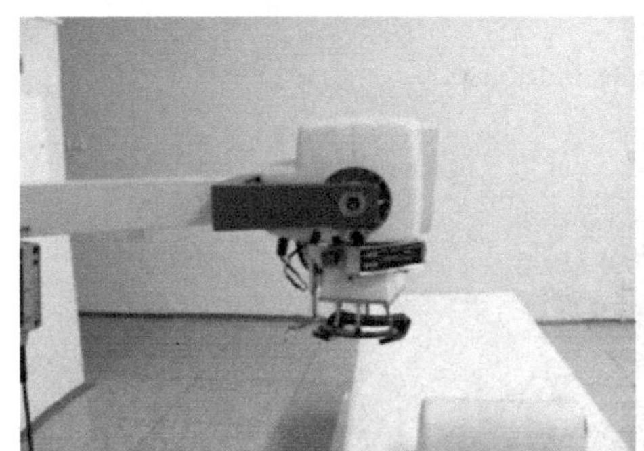

FIG. XVIII–1. Devices containing DU, including a teletherapy head, GammaMat camera and transport container (left to right).

FIG. XVIII–2. National radioactive waste repository at Mochovce.

On 30 November 2011, the Slovak Cabinet approved an action plan between the Slovak Republic and the USA to improve the Slovak Republic's capabilities to prevent, detect and efficiently respond to attempts at nuclear and radiological smuggling. The action plan provided for security upgrades and regular checks at the country's nuclear installations, updating of the national register of radioactive sources, enhancement of the physical protection of facilities with high activity radioactive sources, and development and improvement of human resources in nuclear security. The cooperation also envisaged the establishment of a central facility for long term storage of abandoned and disused radioactive sources.

XVIII–3. REGULATIONS FOR SAFETY AND SECURITY

There are general safety regulations for the handling of nuclear materials. The authorities responsible for the supervision of DU include:

— The Nuclear Regulatory Authority of the Slovak Republic;
— The Public Health Authority of the Slovak Republic;
— The Ministry of Transport and Construction of the Slovak Republic.

With regard to security, there are no special regulations relating to disused DU shielding. However, there are laws and regulations for authorization for:

— Nuclear materials management at nuclear facilities;
— Imports or exports of nuclear materials.

XVIII–4. EXPERIENCE REGARDING RETURN TO MANUFACTURER/SUPPLIER

There have been several cases of returning DSRSs back to the original manufacturer in the Slovak Republic.

All remaining DU shielding material is currently placed in interim storage.

XVIII–5. EXPERIENCE REGARDING REUSE AND RECYCLING

None.

XVIII–6. HAVE YOU IDENTIFIED ANY PROBLEMS OR NEED FOR ASSISTANCE FROM THE IAEA?

Assistance to find the best solution for the management of DU shielding and improve the existing security of DU shielding material. In addition, assistance and/or guidance is required to extend the life of the GammaMat certificate (shielding), which requires strict compliance with the requirements of the ISO 3999:2004(E) Standard, as well as the validity of the authorization certificates for package models.

ISO 3999:2004(E) specifies the performance, design and test requirements of apparatus for gamma radiography with portable, mobile and fixed exposure containers (such as GammaMats) of the various categories defined in Clause 4.

Annex XIX

MEMBER STATE EXPERIENCE: SLOVENIA

XIX–1. TYPES OF DEVICES THAT CONTAIN DU

The Slovenian inventory that contains DU consists of projectors, transport containers and source changers for industrial radiography, shielding and collimators in teletherapy heads and industrial gauges (Fig. XIX–1).

Manufacturer	Device model	Photo
ISOTOPEN-TECHNIK DR. SAUERWEI, GmbH, Haan, Germany	• GammaMat TI-F • GammaMat TI	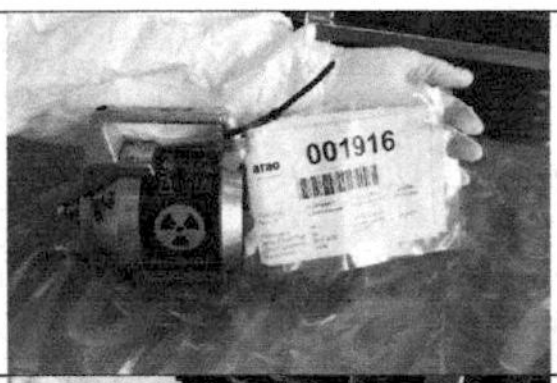
KOWOL GmbH, Düsseldorf, Germany	• Kowomat SU 100, • Kowomat F 100	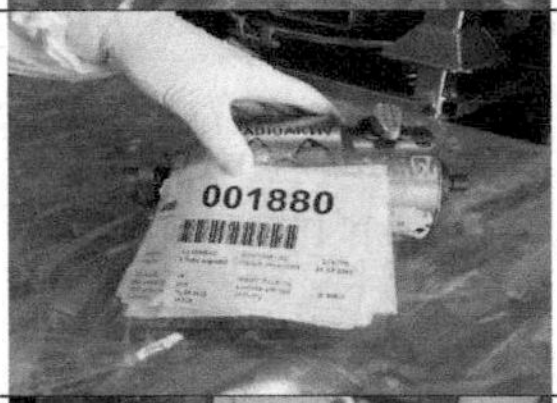
Rich. Seifert & Co. Ahrensburg, Germany	• Gammavolt CO 30	
NUCLEAR GmbH, Düsseldorf, Germany	• TVB	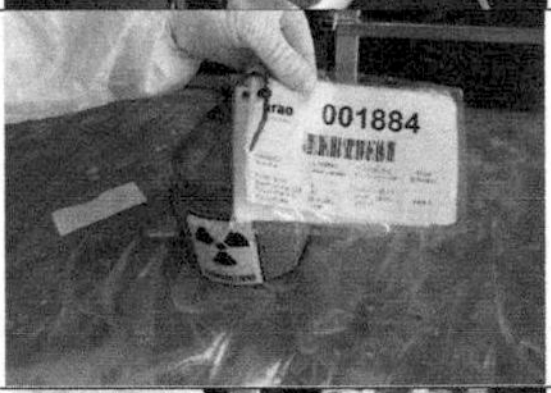
	• Collimator	
THERMO	• Gamma Ray RM M 200 E-M	

FIG. XIX–1. Summary of devices and containers containing DU as shielding material for radiation.

XIX–2. REGULATIONS FOR SAFETY AND SECURITY

A device or container that contains DU is considered as a radioactive source. A licence must be obtained to conduct radiation practices, for the use of a radiation source or for registration of radiation sources. An applicant shall submit a plan for the use and storage of the radiation source, as well as a plan for the handling of the radioactive waste resulting from the radiation practice.

The use of a radiation source that does not comply with the technical documentation is not allowed. Annual inspections of sources (devices/equipment/containers) are required. The authorized radiation protection organizations perform these inspections. There are also regular inspections by the national regulatory authority.

DU is also considered as a nuclear material. The user of nuclear material must keep nuclear material accounting records. The records must be kept in accordance with article 7 of the Commission Regulation (Euratom) No. 302/05, using the application of Euratom Safeguards. The nuclear material inventory must be reported regularly to the national regulatory authority and to the European Commission.

When transporting DU, the shipments must comply with the provisions of the national Dangerous Goods Transport Act (ADR is implemented in the national legislation). For portable devices, the devices or containers must have a valid transport certificate.

During the useful lifetime of the device or the container that contains DU, the user must keep it in temporary storage at their premises. When the device or the container that contains DU is disused, the user has three options:

- — Transfer it to the CSF (see Fig. XIX–2);
- — Return it to manufacturer/supplier;
- — Transfer it to another authorized user.

XIX–3. STORAGE OF DISUSED DU IN THE CSF

Disused DU is either returned to the supplier or shipped to the CSF for institutional radioactive waste. The facility is equipped with a ventilation system for reducing radon concentration and air contamination in the storage facility. To obtain relatively low and constant humidity, it is equipped with an air drying system.

The users of DU must pay a fee for the storage and its further management. The responsibility for accepted DU is transferred from the user to the State.

FIG. XIX–2. Disused DU is stored in the CSF in metal drums or metal box-pallets.

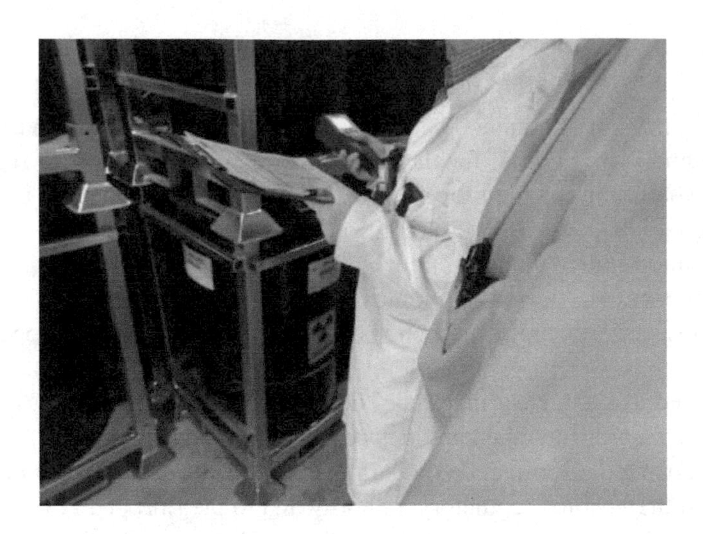

FIG. XIX–3. Euratom Safeguards inspectors inspect nuclear installations every year. Some inspections are undertaken jointly by inspectors from both Euratom and the IAEA.

Currently, the CSF operator is intensifying its efforts to promote sustainable options for the reuse and recycling of DU.

The operator of the CSF keeps records in accordance with the Commission Regulation (Euratom) No. 302/05 in the software application of Euratom Safeguards. The operator of the CSF inspects all batches that contain nuclear material in the CSF at least every calendar year (see Fig. XIX–3). The records on the nuclear material are managed as confidential information. The personnel dealing with these records must pass national security screening.

XIX–4. EXPERIENCE REGARDING RETURN TO MANUFACTURER/SUPPLIER

Return of DU devices to the manufacturer/supplier is an option, but it is not common practice by users. Such returns are usually not free of charge for users. There have been a few cases when the suppliers offered to accept old DU devices free of charge or with a greater discount price. Usually this kind of return is carried out in cases when users purchase a new device from the same supplier.

XIX–5. EXPERIENCE REGARDING REUSE AND RECYCLING

In 2018, the operator of the CSF released two DU containers from the CSF for reuse in a new application (industrial radiography) to a new authorized user.

Recycling of DU is available abroad. It is expensive, and for this reason is not a common practice by users. However, there has been an example when a user ordered the removal of a teletherapy device from Slovenia (Theratron, 110 TBq ^{60}Co, 100 kg of DU shield and 2 t of lead). According to the available information, the ^{60}Co source and DU shield were recycled.

In 2018, approximately 100 kg of DU from the CSF was exported to Czech Republic, where recycling was provided. The total mass of the shipment was 145 kg, and the total mass of DU was approximately 100 kg with activity of 1.8 GBq (see sequence of activities in Figs XIX–4 to XIX–7). More shipments of DU for recycling are planned.

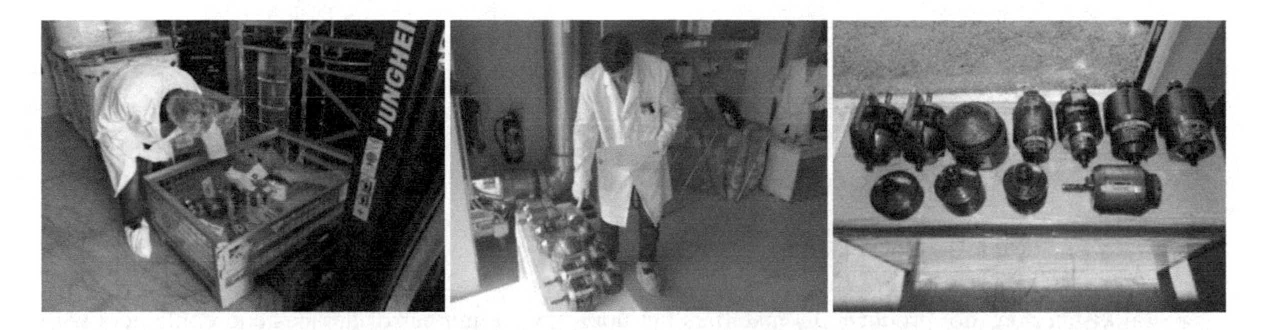

FIG. XIX–4. Selection and preparation of suitable devices/containers for shipment.

FIG. XIX–5. Selected devices/containers were loaded into the vehicle at the CSF site. The licensed transport container (Type A package) was used as an overpack.

FIG. XIX–6. Transfer of DU from Ljubljana to Prague. The final package fulfilled all requirements for an excepted package and was classified under UN 2909 radioactive material, excepted package — articles manufactured from natural uranium or DU or natural thorium.

FIG. XIX–7. At the final destination in Prague, the DU devices/containers were successfully unloaded.

Annex XX

MEMBER STATE EXPERIENCE: TAJIKISTAN

XX–1. TYPES OF DEVICES THAT CONTAIN DU

Tajikistan does not produce DU materials but does have a number of devices and containers with DU as a shielding material, including medical radiotherapy units, industrial level gauges and research irradiators (Fig. XX–1).

XX–2. STORAGE AND DISPOSAL FACILITIES FOR DU

Disused sources are stored under the control of the regulatory authority on the user's premises until they can be shipped to the National Waste Disposal Site, located 50 km from Dushanbe, for long term storage.

XX–3. REGULATIONS FOR SAFETY AND SECURITY

There are safety regulations for the management of this material as a form of radioactive source (storage, transport and disposal), as well as under the IAEA safeguards agreement for all DU materials to be controlled and declared.

Currently, Tajikistan is revising its Law on Radiation Safety to bring it into full compliance with IAEA GSR Part 3. This revised law obliges all registered users to make an agreement with the manufacturer to return imported radioactive sources to the country of origin once they become disused.

FIG. XX–1. Transport containers, industrial gauges and radiotherapy units.

XX–4. EXPERIENCE REGARDING RETURN TO MANUFACTURER/SUPPLIER

None to date.

XX–5. EXPERIENCE REGARDING REUSE AND RECYCLING

None to date.

XX–6. HAVE YOU IDENTIFIED ANY PROBLEMS OR NEED FOR ASSISTANCE FROM THE IAEA?

In the case of teletherapy heads, Tajikistan needs assistance to return the DSRSs to the suppliers or send them to recycling companies to remove the DSRSs for recycling, as well as to return the shielded transport containers back to Tajikistan for reuse.

Annex XXI

MEMBER STATE EXPERIENCE: THAILAND

XXI–1. TYPES OF DEVICES THAT CONTAIN DU

These include gamma radiography cameras, source changers, teletherapy heads and well logging equipment. There are approximately 400 containers and devices containing DU currently in storage around Thailand, with a total mass of approximately 8000 kg of DU. A further 23 items (700 kg mass) is stored in a CSF.

XXI–2. STORAGE AND DISPOSAL FACILITIES FOR DU

Devices in use are stored at the user's premises. Disused DU containers/devices are stored at a CSF at the Thailand Institute of Nuclear Technology (TINT), Fig. XXI–1.

XXI–3. REGULATIONS FOR SAFETY AND SECURITY

Owners of source material must provide the Office for Atoms for Peace (OAP) with the following information as part of their operating licence:

— Amount in possession;
— Purpose(s) of use;

FIG. XXI–1. Centralized DSRS storage facility at TINT.

— Location of use/storage;
— Waste management plan or plan to return to manufacturer;
— If used with a radioactive source, a radiation protection programme and an emergency plan.

In relation to security, the regulation in place for this source material states that the owner of the source material must have an appropriate management plan for security by implementing the following measures:

— Ensuring access control of source material;
— Keeping the material in a locked room or locked container;
— Using nuclear material accounting;
— Reporting to the OAP in the case of lost or stolen materials.

Safeguards regulations require the reporting of nuclear material accounting and balance to the OAP.

XXI–4. EXPERIENCE REGARDING RETURN TO MANUFACTURER/SUPPLIER

DSRS Category 1 and 2 disused sources must be returned to the manufacturer. Obsolete models remain in Thailand. After the licences expire, they must be sent to the central waste storage facility at TINT.

Currently, DU and DU-containing devices are stored at the TINT centralized waste storage facility and/or on the premises of the users.

There is no current plan for disposal of radioactive waste and/or DU.

Identification of orphan or unlabelled DSRS containers for source type and DU shielding is always a challenge, as shown in Fig. XXI–2.

XXI–5. EXPERIENCE REGARDING RETURN TO MANUFACTURER/SUPPLIER

No experience in this area. Interim storage is the preferred and lowest cost option at the current time.

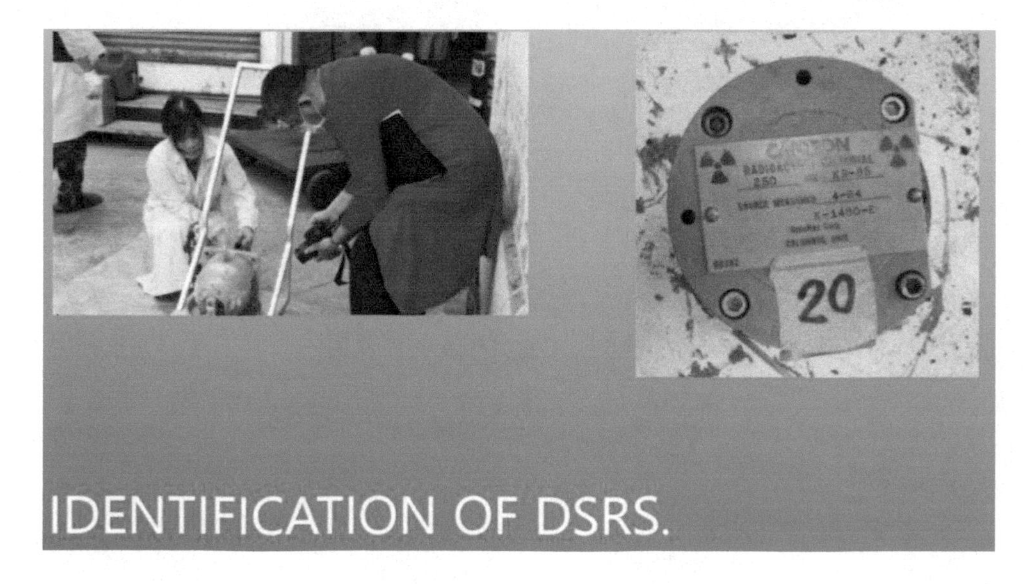

FIG. XXI–2. Identification challenge in orphan DSRS containers/devices.

XXI–6. HAVE YOU IDENTIFIED ANY PROBLEMS OR NEED FOR ASSISTANCE FROM THE IAEA?

For the regulator:

— Guidance on how to maintain an accurate database and tracking system for all nuclear materials;
— Verification/inspection of accounting.
— For the operator:
— Guidance on management of DU from expired models;
— Safety, security and safeguards measures to be put in place for the storage/disposal of DU.

Annex XXII

MEMBER STATE EXPERIENCE: Türkiye

XXII–1. TYPES OF DEVICES THAT CONTAIN DU

Teletherapy heads: 90 teletherapy devices with DU shielding; different models and types are in temporary storage (Fig. XXII–1).

Gammagraphy devices: 157 gammagraphy devices with DU shielding, different models and types; some of them still contain the DSRSs (Fig. XXII–2).

XXII–2. STORAGE AND DISPOSAL FACILITIES FOR DU

In Türkiye, all used devices that become radioactive waste have to be repatriated to the original supplier, or, if this is not possible, they are sent to the radioactive waste management department of the Turkish Atomic Energy Authority (TAEK) and then taken to TAEK's centralized waste processing and storage facility (Fig. XXII–3).

FIG. XXII–1. Teletherapy heads in temporary storage.

FIG. XXII–2. Radiography devices with DU shielding in storage.

FIG. XXII–3. Disused devices containing DU stored at TAEK's centralized waste processing and storage facility.

XXII–3. REGULATIONS FOR SAFETY AND SECURITY

Currently there are no specific safety regulations for containers or devices containing DU shielding. The regulations below cover the safety framework of DU content devices:

— Radioactive Waste Management Regulation (09.03.2013);
— Regulation on the Control of High Activity Sealed Radioactive Sources and Orphan Sources (21.03.2009);
— Regulation on Radiation Protection and Licensing on Industrial Radiography Devices (08.07.2005).

Türkiye has no direct experience regarding the security of DU shielding materials used in transport containers or devices. However, the regulations below provide the framework for security of related materials.

— Regulation on Radiation Safety, 2000;
— Regulation on Nuclear and Radiological National Emergency Preparedness, 2000.

XXII–4. EXPERIENCE REGARDING RETURN TO MANUFACTURER/SUPPLIER

Repatriation of DU-containing transport containers and devices is governed by the nuclear regulatory authority (NDK). TAEK's radioactive waste management department transports, receives and stores all radioactive waste (including DU) within Türkiye.

XXII–5. EXPERIENCE REGARDING REUSE AND RECYCLING

Türkiye has not considered recycling at this stage, with interim storage being the preferred choice.

XXII–6. HAVE YOU IDENTIFIED ANY PROBLEMS OR NEED FOR ASSISTANCE FROM THE IAEA?

For Türkiye:

- TAEK's radioactive waste management department shares information on DU content with the NDK, which send the reports to the IAEA for Nuclear Safeguards Reports and nuclear material accounting procedures.
- Some devices do not have labelling or documentation on DU content or other useful information. Hence there is a need for a technical specific approach to assist with such unknown DU-containing devices.
- The advantage of having the same brand or model as other similar devices provides the best evidence and information for identifying such unknown devices.

For potential DU-containing unlabelled containers (with a ^{60}Co source inside), the gamma spectrometry measurement and identification method was used at the TAEK Radioactive Waste Management Centre but still could not identify if DU was present. This issue needs to be resolved.

Annex XXIII

MEMBER STATE EXPERIENCE: UNITED STATES OF AMERICA

XXIII–1. TYPES OF DEVICES THAT CONTAIN DU

The USA has an extensive amount of devices and equipment that use DU as shielding. This includes radiography cameras, Type A and Type B medical casks, gauges, irradiators using ^{60}Co sources, scientific devices, tritium getters, targets for medical isotopes, multipliers, collimators, teletherapy units, and so on.

XXIII–2. STORAGE AND DISPOSAL FACILITIES FOR DU

The USA does not have a CSF for DU shielding. Industrial and commercial users store DU shielding at their facilities. The Department of Energy manages the National Laboratories, which store a considerable amount of DU shielding.

There are three options for burial disposal in the USA. Material can be sent to Energy Solutions in Clive, Utah; WCS in Andrews, Texas; or the Nevada National Security Site (previously the Nevada Test Site) in Nye County, Nevada. The Nevada National Security Site is operated for the United States Department of Energy. Burial disposal is not an option for material of non-domestic (outside the USA) origin.

XXIII–3. REGULATIONS FOR SAFETY AND SECURITY

The USNRC regulates the possession, use and disposal of DU. The regulations are administered by the USNRC or by Agreement States. Agreement States have entered into agreements with the USNRC that give them the authority to licence and inspect by-products, sources or special nuclear materials used or possessed within their borders.

Manufacturing Sciences Corporation (MSC) in Oak Ridge, Tennessee, is licensed by the State of Tennessee to possess up to 874 042 kg of DU and 115 666 kg of natural uranium. It reports all radioactive material transfers and receipts through the Nuclear Materials Management and Safeguards System. This system is designed to inform and advance US Government policy and nuclear material accounting related to domestic and international safeguards, non-proliferation, national security and global commerce for peaceful uses of nuclear material.

All material accepted into MSC inventory is MSC's property and responsibility. The MSC facility is located on 19 fenced acres and has controlled secure access. MSC has an excellent relationship with the State of Tennessee regulators, which enforce USNRC security regulations. Worker safety is also regulated by the USNRC and the State of Tennessee.

XXIII–4. DU SHIELDING AT MSC

MSC has shipped DU shielding from 2 kg to 4000 kg, both within the USA and internationally. MSC has returned DU shielding to customers who have had it processed at MSC for their use. MSC is a manufacturing facility with casting capabilities (Figs XXIII–1 to XXIII–3).

MSC has provided the following DU shielding services:

— Rolling services of plate, sheet and foil products;

FIG. XXIII–1. A 4500 kg vacuum induction melting casting machine with plate and foil rolling capabilities.

FIG. XXIII–2. Four high reversing rolling mill. Able to machine DU components (ranging from 1 kg to 4000 kg).

FIG. XXIII–3. Machined DU spheres (left) and a large machined DU shielded teletherapy head (right).

- — Shields and casks for DU sources;
- — DU products for radiation shields;
- — Forming and machining of DU components used in DU shielding applications;
- — DU speciality shapes for specific applications;
- — Recycling of DU and natural uranium into new DU shielding components;
- — Recycling of DU and natural uranium into products to service existing business lines;
- — Process development;
- — Pilot production development for scale-up;
- — Medical isotope targets.

MSC has been a leader in the production of speciality DU shields. It has done the following:

- — Produced the largest DU casting in North America;
- — Developed advanced rolling practices for DU sheet using in-line infrared reheating technology, resulting in a significant reduction of process time and a fine grained formable 0.75 mm thick sheet;
- — Cast, rolled and machined precision penetrometer foil components as thin as 0.025 mm;
- — Cast and machined thousands of DU shielded medical isotope casks;
- — Cast and machined DU camera bodies for field radiography;
- — Cast and machined 1600 kg DU shields for ^{60}Co sources used in radiation therapy.

XXIII–5. EXPERIENCE REGARDING REUSE AND RECYCLING

MSC accepts DU shielding from domestic and international customers for recycling. MSC has recycled > 1 000 000 kg of DU shielding material into new DU products for its customers.

MSC has recycled teletherapy heads from South America for the IAEA (Figs XXIII–4 and XXIII–5).

MSC was part of a team that returned ten teletherapy heads to the USA. A company moved a portable hot cell to the site, removed the ^{60}Co sources and shipped the heads to the USA, where the packages were disassembled. MSC recycled the DU. The ^{60}Co and the steel components were also recycled.

XXIII–6. HAVE YOU IDENTIFIED ANY PROBLEMS OR NEED FOR ASSISTANCE FROM THE IAEA?

No.

FIG. XXIII–4. Siemens Gammatron head.

FIG. XXIII–5. Alcyon head.

Annex XXIV

MEMBER STATE EXPERIENCE: VIET NAM

XXIV–1. TYPES OF DEVICES THAT CONTAIN DU

Devices are predominantly radiography cameras (Fig. XXIV–1). However, Viet Nam has over 20 ^{60}Co teletherapy units (some containing DU shielding), with some of these units and sources having been supplied during the 1990s through international aid donation. These sources, which have now decayed, are no longer of clinical use and remain in temporary storage at hospitals around the country.

XXIV–2. STORAGE AND DISPOSAL FACILITIES FOR DU

The DSRSs and DU devices are stored in several storage facilities located at the Da Lat Nuclear Research Institute (DNRI) and the Institute of Nuclear Science and Technology — see Fig. XXIV–2. However, Viet Nam does not have a single CSF for radioactive sources or other nuclear materials such as containers or devices containing DU shielding. A project to establish a CSF is currently under consideration.

FIG. XXIV–1. Radiography cameras (some with DU shielding) in interim storage.

FIG. XXIV–2. Storage facility at DNRI for disused radioactive sources and radiography cameras.

XXIV–3. REGULATIONS FOR SAFETY AND SECURITY

In the last few years, several laws and regulations on radiation safety and radioactive source security including DU have been produced and are being implemented. Viet Nam is thus recognizing and addressing its legacy source and radioactive waste issues.

The Viet Nam Agency for Radiation and Nuclear Safety (VARANS) has expanded to VARANSAC, which includes the control (including security regulations required) to safely and securely store the disused sources and associated DU waste containers.

XXIV–4. EXPERIENCE REGARDING RETURN TO MANUFACTURER/SUPPLIER

Viet Nam has some experience with interim storage of disused radioactive sources and devices/containers used to transport the sources. Some of the used containers have DU material used for shielding.

VARANS has provided guidance when an SRS is no longer to be used for its dedicated purpose. The following management options may be considered:

- Transfer to another user for application elsewhere;
- Return to the manufacturer/supplier;
- Storage for decay of sources containing radionuclides with short half-life, followed by discharge as non-radioactive material;
- Transportation to a centralized interim storage facility until a conditioning facility is available;
- Transportation to a central conditioning facility for conditioning, followed by interim storage.

XXIV–5. EXPERIENCE REGARDING REUSE AND RECYCLING

Viet Nam has no experience in reuse or recycling — only with interim storage. Also, there has not been any significant active recovery of the sources by the supplying countries. The small inventory of DU material in Viet Nam has not warranted recycling at this stage.

XXIV–6. HAVE YOU IDENTIFIED ANY PROBLEMS OR NEED FOR ASSISTANCE FROM THE IAEA?

Yes. Further support in terms of training on nuclear materials and the disposition pathways for DU materials arising from radioactive source containers and devices is required in the future.

GLOSSARY

The following terms are used in this publication, as defined here.

Unless noted specifically, all definitions are extracted either from the IAEA Safety Glossary[1], marked with *, or from the IAEA Safeguards Glossary[2], marked with **.

Agreement State. In the USA, Agreement States have entered into agreements with the USNRC that give them the authority to licence and inspect by-products, sources or special nuclear materials used or possessed within their borders.[3]

ALARA*. Optimization (of protection and safety). The process of determining what level of protection and safety would result in the magnitude of individual doses, the number of individuals (workers and members of the public) subject to exposure and the likelihood of exposure being 'as low as reasonably achievable', economic and social factors being taken into account.

brachytherapy. A type of radiation therapy in which radioactive material sealed in small sealed sources, needles, seeds, wires or catheters is placed directly into or near a tumour.[4]

competent authority*. Any body or authority designated or otherwise recognized as such for any purpose in connection with the Regulations for the Safe Transport of Radioactive Material. Otherwise, the more general term regulatory body should be used, with which competent authority is essentially synonymous

contamination*. Radioactive substances on surfaces or within solids, liquids or gases (including the human body), where their presence is unintended or undesirable, or the process giving rise to their presence in such places.

According to SSR-6 (Rev. 1) contamination means the presence of a radioactive substance on a surface in quantities in excess of 0.4 Bq/cm^2 for beta and gamma emitters and low toxicity alpha emitters, or 0.04 Bq/cm^2 for all other alpha emitters.

cost–benefit analysis*. A systematic technical and economic evaluation of the positive effects (benefits) and negative effects (disbenefits, including monetary costs) of undertaking an action.

decommissioning*. Administrative and technical actions taken to allow the removal of some or all of the regulatory controls from a facility.

depleted uranium*. Uranium containing a lesser mass percentage of ^{235}U than is present in natural uranium.

---

[1] INTERNATIONAL ATOMIC ENERGY AGENCY, IAEA Safety Glossary, Terminology Used in Nuclear Safety and Radiation Protection, 2018 Edition, IAEA, Vienna (2019).

[2] INTERNATIONAL ATOMIC ENERGY AGENCY, IAEA Safeguards Glossary, International Nuclear Verification Series No. 3, 2001 Edition, IAEA, Vienna (2002).

[3] US NUCLEAR REGULATORY COMMISSION, Agreement State Program, USNRC, Washington, DC, https://www.nrc.gov/about-nrc/state-tribal/agreement-states.html.

[4] NATIONAL CANCER INSTITUTE, Dictionary of Cancer Terms, NCI, Rockville, MD, https://www.cancer.gov/publications/dictionaries/cancer-terms/def/brachytherapy

device. In this publication, a device holds a radioactive source for use in its given application. It provides radiation shielding and may also allow a controlled beam of radiation to be used for a desired purpose. Devices include:

— Teletherapy heads: primary shielding, collimators and trimmer bars;
— Brachytherapy afterloader and source changing and storage units with DU shielding;
— Self-shielded irradiators (e.g. blood and laboratory irradiators);
— Exposure devices and collimators used in industrial gamma radiography;
— Industrial gauges (level, flow, density, etc.) shielded with DU;
— DU shielding blocks used in well logging tools;
— Transport packages and storage containers containing DU shielding.

disposal*. Emplacement of waste in an appropriate facility without the intention of retrieval.

disused source*. A radioactive source that is no longer used, and is not intended to be used, for the practice for which an authorization has been granted (see Code of Conduct on the Safety and Security of Radioactive Sources).

Note: A disused source may still represent a significant radiological hazard. It differs from a spent source in that it may still be capable of performing its function; it may be disused because it is no longer needed.

exemption (1)*. The determination by a regulatory body that a source or practice need not be subject to some or all aspects of regulatory control on the basis that the exposure and the potential exposure due to the source or practice are too small to warrant the application of those aspects or that this is the optimum option for protection irrespective of the actual level of the doses or risks. Note that this definition is based only on radiological aspects (see exemption (2)).

exemption (2).** The "exemption of nuclear material from safeguards on account of its use or quantity" is addressed in points 2.13, 6.14 and 6.24 of the IAEA Safeguards Glossary. Also applicable is the concept of termination.

exposure device. In industrial gamma radiography, the exposure device contains the SRS, which will be driven by a cable to the working position.[5] In this publication, the exposure device is sometimes called a 'projector' or 'camera'.

hazard*. The potential for harm or other detriment; a factor or condition that might operate against safety. Note that this publication makes a distinction between radiological and chemical/industrial risks.

intermediate level radioactive waste (ILW)*. Radioactive waste that, because of its content, in particular its content of long lived radionuclides, requires a greater degree of containment and isolation than that provided by near surface disposal. Note that for the purposes of this publication, 'waste' means 'disused DU device'.

irradiator (irradiation installation)*. A structure or an installation that houses a particle accelerator, X ray apparatus or large activity radioactive source and that can produce high radiation fields.

[5] INTERNATIONAL ORGANIZATION FOR STANDARDIZATION, ISO 3999:2004 "Radiation protection — Apparatus for industrial gamma radiography — Specifications for performance, design and tests", ISO, Geneva (2004).

linear accelerator (or linac). A device in which electrons are accelerated in a straight line by successive impulses from a series of electric fields to produce electron beams or, after conversion in a dense target, high fluxes of X rays to the region of the patient's tumor for teletherapy purposes.

licence*. Legal document issued by the regulatory body granting authorization to perform specified activities relating to a facility or activity.

licensee*. The holder of a current licence. The licensee is the person or organization having overall responsibility for a facility or activity.

low level radioactive waste (LLW)*. Radioactive waste that is above clearance levels, but with limited amounts of long lived radionuclides. Note that for the purposes of this publication, 'waste' means 'disused DU device'.

management (of SRSs)*. The administrative and operational activities that are involved in the manufacture, supply, receipt, possession, storage, use, transfer, import, export, transport, maintenance, recycling or disposal of radioactive sources.

This usage is specific to the Code of Conduct on the Safety and Security of Radioactive Sources.[6]

operator (operating organization)*. Any person or organization applying for authorization, or authorized and/or responsible for safety, when undertaking activities in relation to any nuclear facilities or sources of ionizing radiation.

overpack*. A secondary (or additional) outer container for one or more waste packages, used for handling, transport, storage and/or disposal. Note that for the purposes of this publication, 'waste' means 'disused DU device'.

According to SSR-6 (Rev. 1), overpack means an enclosure used by a single consignor to contain one or more packages and to form one unit for convenience of handling and stowage during transport.

packaging*. Preparation of radioactive waste for safe handling, transport, storage and/or disposal by means of enclosing it in a suitable container. Note that for the purposes of this publication, 'waste' means 'disused DU device'.

According to SSR-6 (Rev. 1), packaging means one or more receptacles and any other components or materials necessary for the receptacles to perform containment and other safety functions.

radioactive source*. Radioactive material that is permanently sealed in a capsule, or closely bonded and in a solid form, and is not exempt from regulatory control.

This definition is particular to the Code of Conduct on the Safety and Security of Radioactive Sources.[7]

radioactive waste*. For legal and regulatory purposes, material for which no further use is foreseen that contains, or is contaminated with, radionuclides at activity concentrations greater than clearance levels as established by the regulatory body.

---

[6] INTERNATIONAL ATOMIC ENERGY AGENCY, Code of Conduct on the Safety and Security of Radioactive Sources, IAEA/CODEOC/2004, IAEA, Vienna (2004).

[7] INTERNATIONAL ATOMIC ENERGY AGENCY, Code of Conduct on the Safety and Security of Radioactive Sources, IAEA/CODEOC/2004, IAEA, Vienna (2004).

recycling*. The process of converting waste materials into new products. Note that for the purposes of this publication, 'waste' means 'disused DU device'.

regulatory body*. An authority or a system of authorities designated by the government of a State as having legal authority for conducting the regulatory process, including issuing authorizations, and thereby regulating nuclear, radiation, radioactive waste and transport safety.

reuse*. The use of an item again after it has been used before.

(coverage of IAEA) safeguards.** The scope of application defined by the relevant safeguards agreement. Under a comprehensive safeguards agreement (CSA), safeguards are applied on "all source or special fissionable material in all peaceful nuclear activities within the territory of [the] State, under its jurisdiction, or carried out under its control anywhere." Thus, such agreements are considered comprehensive (or 'full scope'). The scope of a CSA is not limited to the nuclear material declared by a State but includes all nuclear material subject to IAEA safeguards.

self-shielded irradiator. A irradiator in which the sealed source(s) is completely contained in a dry container constructed of solid materials. The sealed source(s) is shielded at all times, and human access to the sealed source(s) and the volume undergoing irradiation is not physically possible in its designed configuration[8].

(radiation) shielding. Physical barriers designed to provide protection from the effects of ionizing radiation.

source material.** This is defined as "uranium containing the mixture of isotopes occurring in nature; uranium depleted in the isotope 235; thorium; any of the foregoing in the form of metal, alloy, chemical compound, or concentrate; any other material containing one or more of the foregoing in such concentration as the Board of Governors shall from time to time determine; and such other material as the Board of Governors shall from time to time determine".

storage*. The holding of radioactive sources, radioactive material, spent fuel or radioactive waste in a facility that provides for its containment, with the intention of retrieval.

teletherapy. The treatment of diseased tissue with high intensity radiation (as gamma rays from radioactive cobalt)[9].

termination of IAEA safeguards.** Safeguards in a given State normally continue on nuclear material (and subsequent generations of nuclear material produced therefrom) until the material is transferred to another State that has assumed responsibility for it, or until the material has been consumed or has been diluted in such a way that it is no longer usable for any nuclear activity relevant from the point of view of safeguards, or has become practicably irrecoverable. This point is addressed in points 2.12, 6.14 and 6.25 of the IAEA Safeguards Glossary.

transport (transportation)*. The deliberate physical movement of radioactive material (other than that forming part of the means of propulsion) from one place to another. See also paragraphs 106 and 107 of SSR-6 (Rev. 1) [29].

---

[8] US NUCLEAR REGULATORY COMMISSION, Consolidated Guidance About Materials Licenses Program — Specific Guidance About Self-Shielded Irradiator Licenses — Final Report, NUREG-1556, Vol. 5, Rev. 1, USNRC, Washington, DC (2018).

[9] See: https://www.merriam-webster.com/medical/teletherapy

very low level waste (VLLW)*. Radioactive waste that does not necessarily meet the criteria of exempt waste, but that does not need a high level of containment and isolation and, therefore, is suitable for disposal in landfill-type near surface repositories with limited regulatory control.

waste acceptance criteria*. Quantitative or qualitative criteria specified by the regulatory body, or specified by an operator and approved by the regulatory body, for the waste form and waste package to be accepted by the operator of a waste management facility.

waste container*. The vessel into which the waste form is placed for handling, transport, storage and/or eventual disposal; also, the outer barrier protecting the waste from external intrusions. Note that for the purposes of this publication, 'waste' means 'disused DU device'.

ABBREVIATIONS

ADR	European Agreement Concerning the International Carriage of Dangerous Goods by Road
BSS	Basic Safety Standards
CFR	Code of Federal Regulations (USA)
CSA	comprehensive safeguards agreement
CSF	centralized storage facility
DSRS	disused sealed radioactive source
DU	depleted uranium
Euratom	European Atomic Energy Community
HDR	high dose rate
HLW	high level waste
ILW	intermediate level waste
LDR	low dose rate
LLW	low level waste
MDR	medium dose rate
NPT	Treaty on the Non-Proliferation of Nuclear Weapons
SRS	sealed radioactive source
USNRC	US Nuclear Regulatory Commission
VLLW	very low level waste

CONTRIBUTORS TO DRAFTING AND REVIEW

Abdalrahman, A.	Jordan Atomic Energy Commission, Jordan
Armozd, H.R.	Atomic Energy Organization of Iran, Islamic Republic of Iran
Basilia, G.	LEPL Agency of Nuclear and Radiation Safety, Georgia
Benitez-Navarro, J.C.	International Atomic Energy Agency
Boboyorov, M.	Nuclear and Radiation Safety Agency, Tajikistan
Csullog, G.	Canada
Dayal, R.	Canada
Dimitrovski, L.	Australia
Din, S.U.	Pakistan Atomic Energy Commission, Pakistan
Dumitrescu, N.	National Commission for Nuclear Activities Control, Romania
Griffin, J.	Los Alamos National Laboratory, USA
Hageman, J.	Southwest Research Institute, USA
Heard, R.	consultant, South Africa
Hubalek, J.	Gamma-Services, Germany
Ibrahim, M.Z.B.	Malaysian Nuclear Agency, Malaysia
Ivanovic, T.	Nuclear Regulatory Authority of the Slovak Republic, Slovakia
Katona, T.	Hungarian Atomic Energy Authority, Hungary
Koteng, A.O.	Radiation Protection Board, Kenya
Krupička, P.	UJP PRAHA, Czech Republic
Krutzik, R.B.	Manufacturing Sciences Corporation, USA
Laraia, M.	Italy
Levey, T.	Acuren Group, Canada
Mathews, C.	International Atomic Energy Agency
Ndao, A.	Cheikh Anta Diop University, Senegal
Nguyen Trong, H.	Vietnam Atomic Energy Institute, Viet Nam
Nugroho, D.H.	Nuclear Energy Regulatory Agency, Indonesia
Pashayev, R.	Nuclear and Radiological Activity Regulations Agency, Azerbaijan
Pavenayotin, N.	Office of Atoms for Peace, Thailand
Pereira Campos, V.A.	Comision Chilena de Energia Nuclear, Chile

Pillette-Cousin, L.	France
Randriantseheno, H.F.	Institut National des Sciences et Techniques Nucléaires, Madagascar
Reber E.H.	International Atomic Energy Agency
Smith, N.A.	International Atomic Energy Agency
Sučić, S.	Agency for Radwaste Management, Slovenia
Suseanu, I.	International Atomic Energy Agency
Todorov, N.	Nuclear Regulatory Agency, Bulgaria
Yahyaouy, Z.	Atomic Energy Commission, France
Yilmaz, O.	Cekmece Nuclear Research and Training Centre, Türkiye

Technical Meetings

Vienna, Austria: 19–23 August 2019

Consultants Meetings

Vienna, Austria: 5–9 May 2014, 16–20 March 2015,

1–5 February 2016, 18–22 November 2019

Structure of the IAEA Nuclear Energy Series*

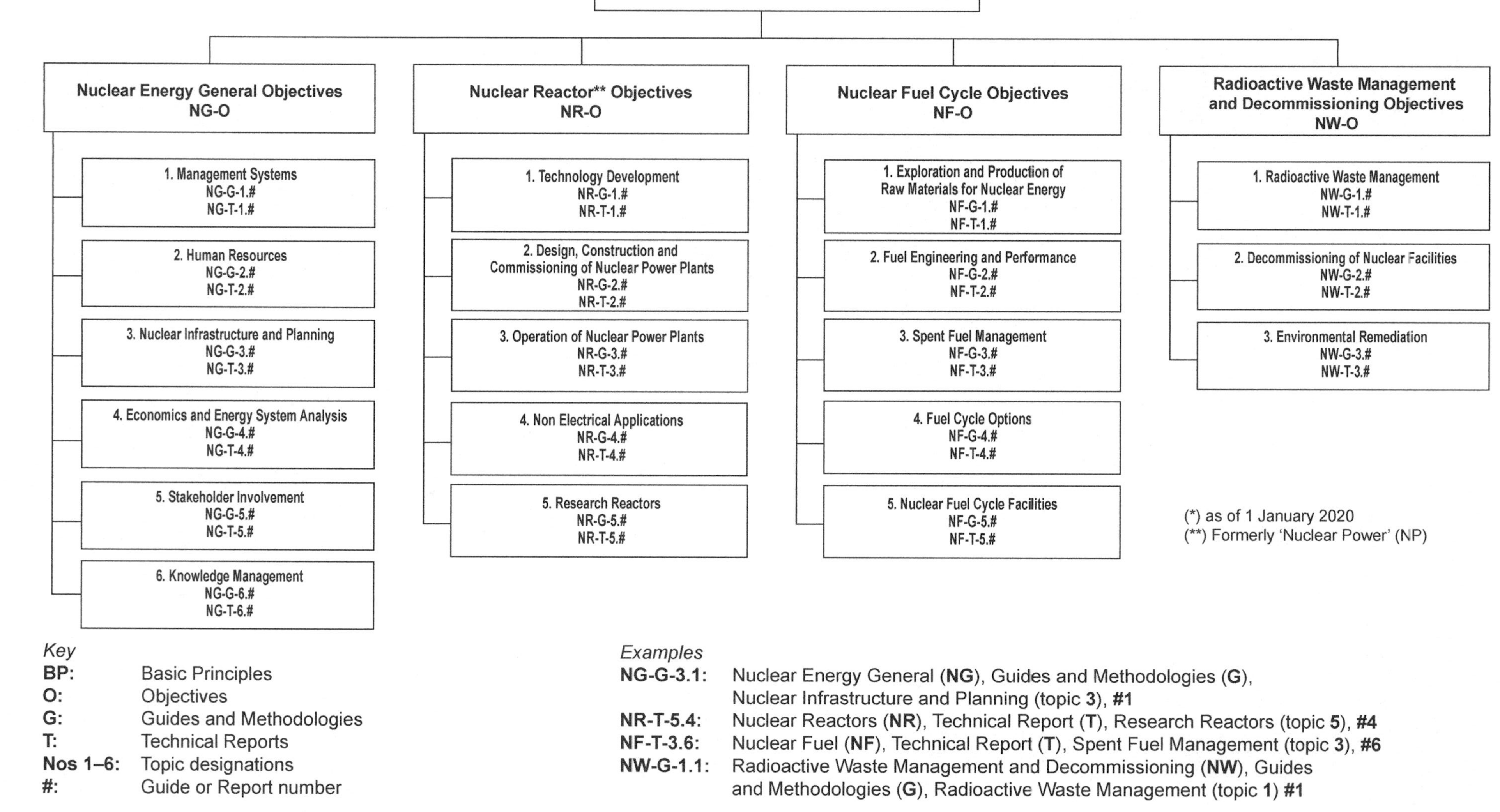

No. 26

ORDERING LOCALLY

IAEA priced publications may be purchased from the sources listed below or from major local booksellers.

Orders for unpriced publications should be made directly to the IAEA. The contact details are given at the end of this list.

NORTH AMERICA

Bernan / Rowman & Littlefield

15250 NBN Way, Blue Ridge Summit, PA 17214, USA
Telephone: +1 800 462 6420 • Fax: +1 800 338 4550
Email: orders@rowman.com • Web site: www.rowman.com/bernan

REST OF WORLD

Please contact your preferred local supplier, or our lead distributor:

Eurospan Group

Gray's Inn House
127 Clerkenwell Road
London EC1R 5DB
United Kingdom

Trade orders and enquiries:

Telephone: +44 (0)176 760 4972 • Fax: +44 (0)176 760 1640
Email: eurospan@turpin-distribution.com

Individual orders:

www.eurospanbookstore.com/iaea

For further information:

Telephone: +44 (0)207 240 0856 • Fax: +44 (0)207 379 0609
Email: info@eurospangroup.com • Web site: www.eurospangroup.com

Orders for both priced and unpriced publications may be addressed directly to:

Marketing and Sales Unit
International Atomic Energy Agency
Vienna International Centre, PO Box 100, 1400 Vienna, Austria
Telephone: +43 1 2600 22529 or 22530 • Fax: +43 1 26007 22529
Email: sales.publications@iaea.org • Web site: www.iaea.org/publications